DESIGN TALENTS

iF concept design award yearbook 2012

IMPRINT

Editor iF DESIGN TALENTS GmbH, Bahnhofstrasse 8, 30159 Hannover, Germany, www.ifdesign.de, phone +49.511.54224-0, fax +49.511.54224-100, info@ifdesign.de ▪ **Managing Director** Ralph Wiegmann ▪ **Vice Director** Rainer Schwarz ▪ **iF International Forum Design GmbH** Bahnhofstrasse 8, 30159 Hannover, Germany and Rosenheimer Str. 145e, 81671 München, Germany ▪ **iF concept design award 2012** Anna Reissert ▪ **iF concept design award yearbook 2012** Anna Reissert, Heike Meier ▪ **iF Press Officer** Annegret Wulf-Pippig ▪ **iF Branch Office Taiwan** 3F., Bldg. G, 3-1 Park Street, Nangang Taipei 115, Taiwan ▪ **iF Representative Office Korea** Designhouse Inc., Taekwang Building, 162-1 Jangchung-Dong 2-GA, Jung-Gu, Seoul 100-855, Korea ▪ **iF Representative Office Brasil** Centro de Design Paraná, Av. Comendador Franco, 1341 - Jardim Botânico, CEP: 80215-090 Curitiba - Paraná, Brasil ▪ **iF Press Office** Claudia Neumann Communication GmbH, Eigelstein 103-113, 50668 Köln, Germany, phone: +49.221.9139490, fax: +49.221.91394919, iF@neumann-luz.de ▪ **iF DESIGN MEDIA GmbH** Ramona Rockel, phone +49.721.35455800, ramona.rockel@ifdesign.de ▪ **Translation and Copy Editing** Sandra Johnson, Hildesheim, Germany ▪ **Corporate Design** helke brandt communication, Hannover, Germany ▪ **Photography** Roman Thomas, Celle, Germany ▪ **Typesetting and Lithograph** Ruth Printmedien GmbH, Braunschweig, Germany ▪ **Print** Offizin Andersen Nexö Leipzig GmbH, Leipzig, Germany

Bibliographic information published by the Deutsche Nationalbibliothek. The Deutsche Nationalbibliothek lists this publication in the Deutsche Nationalbibliografie; detailed bibliografic data are available on the internet at http://dnb.d-nb.de

Partner/Distribution Prestel Verlag, Munich, A member of Verlagsgruppe Random House GmbH ▪ Prestel Verlag Neumarkter Strasse 28, 81673 Munich, phone +49.89.4136-0, fax +49.89.4136-2335, www.prestel.de ▪ Prestel Publishing Ltd. 4 Bloomsbury Place, London WC1A 2QA, phone +44.20.7323-5004, fax +44.20.7636-8004 ▪ Prestel Publishing 900 Broadway, Suite 603, New York, NY 10003, phone +1.212.995-2720, fax +1.212.995-2733, www.prestel.com

Prestel books are available worldwide. Please contact your nearest bookseller or one of the above addresses for information concerning your local distributor.
The Library of Congress Cataloguing-in-Publication data is available. British Library Cataloguing-in-Publication Data: a catalogue record for this book is available from the British Library.

Publisher
iF DESIGN MEDIA GmbH, Bahnhofstrasse 8, 30159 Hannover, Germany, www.ifdesign.de

Printed in Germany
ISBN: 978-3-7913-4705-9

CONTENT

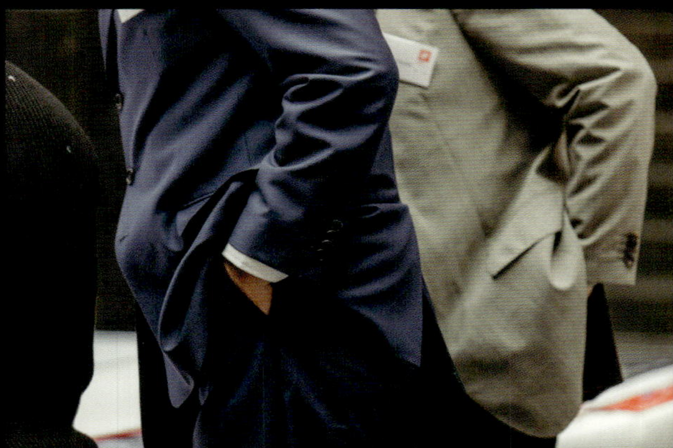

DISCUSSION AND INTUITION

SIGNIFICANT GESTURES

RUSTLING LEAVES

IMPRESSIVE SETTING

THE FINAL VOTE

iF CONCEPT DESIGN AWARD 2012

Promoting young talent is more important today than ever before. Companies need creative minds not only to reflect the individualization of our society, but to shape it and be successful in doing so. Social changes as well as the increasing frequency of natural disasters and the technological challenges posed by such topics as energy and sustainability require outstanding skills and abilities in the natural sciences. The products, processes, procedures and methods that result have to be safe, as long-lasting as possible and easy and intuitive in their operation in professional as well as private environments. A challenge that can be best met only by talented designers!

In the meantime, while creativity transcends geographical borders, it is still subject to cultural differences and other influencing factors. This is very clearly reflected in this year's iF concept design award: 15,000 entries received from students from 62 countries from around the world (of which close to 11,000 were accepted!) provided an exceptional outlook on the products, the communications and the architecture of the future. It became impressively apparent how much the young designers feel inspired by natural disasters, the demographic shift, new energies and other important basic topics and how much they think about ways and approaches to providing the urgently needed solutions to the problems.

Young designers need to gain international experience early in order to be able to fully exploit their design options for their later employers or as self-employed designers.
iF DESIGN TALENTS GmbH, the organizer of this competition, is dedicated to promoting young design talent and with that, the shaping of our future. With the help of supporters and sponsors, we provide design students from all over the world with an important international networking platform.

We would like to express our sincerest thanks to the GOLD sponsors – CeBIT 2012 in Hannover (Germany), Hansgrohe SE (Schiltach, Germany), LG Hausys (Seoul, South Korea), Samsung Electronics Co., Ltd. (Seoul, South Korea) and Volkswagen AG (Wolfsburg, Germany) – as well as to the BRONZE sponsor Coway (Seoul, South Korea), our partner designxport Hamburg (Germany) and LG Electronics, Inc. (Seoul, South Korea) as general supporter – with whose valuable support and commitment we have been able to develop the iF concept design award into one of the most important international design competitions for young designers.

The "Hansgrohe Prize 2012: Efficient Water Design", which was awarded for the second time this year, has also given this competition an important new direction.

Our special thanks also go out to the 56 international designers, professors and lecturers who acted as judges! Of the 10,665 competition entries, they first selected 300 finalists and then filtered out the "Best 100" in a second round, the winners of the iF concept design award 2012. Finally, a total purse of EUR 30,000 was shared between the 20 most outstanding works.

And now it's "your turn" – give these young designers a moment of your time and your attention. Take a look at what the future has in store for us all.

YOURS
RALPH WIEGMANN
iF MANAGING DIRECTOR

JURY FIRST JUDGING ROUND iF CONCEPT DESIGN AWARD 2012

ELAINE ANN
Kaizor Innovation
Hong Kong

SYBS BAUER
designkunst
Hamburg, Germany

PROF. WERNER BAUMHAKL
Hochschule für Gestaltung und
Kunst Basel FNHW
Basel, Switzerland

LI BEI
FIYTA Holdings Ltd.
Shenzhen, China

PROF. RUTH BERKTOLD
Yes Architecture
Munich, Germany

ANNEMARIE BOUTIN
l'agence pour la promotion de la
création industrielle (APCI)
Paris, France

ANGELA CARVALHO
ncs design
Rio de Janeiro, Brasil

HUN-JUNG CHOI
Coway
Seoul, South Korea

AMY Y.M. CHOW
Hong Kong Design Centre
Hong Kong

J'AIME COHEN
The School of Visual Arts /
Masters in Branding Program
New York, United States of America

PROF. CLAUS-CHRISTIAN ECKHARDT
Lund University School of
Industrial Design
Lund, Sweden

ANDREAS ENSLIN
Miele & Cie
Gütersloh, Germany

PROF. ALPAY ER
Istanbul Technical University (ITU)
Istanbul, Turkey

NIKLAS FRINGS-RUPP
Miami Ad School Europe
Hamburg, Berlin, Germany

LETICIA CASTRO GAZIRI
Paraná Design Center (CDPR)
Curitiba, Paraná, Brasil

HUBERT GROTHAUS
DC Design Company
Agentur GmbH
Munich, Germany

FRED HELD
Held + Team
Hamburg, Germany

JEFFREY HO
DesignSingapore Council
Singapore

JAMES IRVINE
James Irvine Srl
Milan, Italy

WILLIAM WU
Haier
Qingdao, China

ZLATKO KAPETANOVIC
University of Zagreb
Zagreb, Croatia

ZEYNEP KARAHAN
Miami Ad School Istanbul
Istanbul, Turkey

ARHAN KAYAR
dDf
Istanbul, Turkey

PROF. MARTIN KÖTTERING
Hochschule für bildende Künste
Hamburg
Hamburg, Germany

JURY FIRST JUDGING ROUND iF CONCEPT DESIGN AWARD 2012

TIAGO KRUSSE
DESIGN MAGAZINE
Lisbon, Portugal

PETER KUCHNICKI
Spark Design Awards
New York, United States of America

JULIAN LINDLEY
University of Hertfordshire
Hertfordshire, United Kingdom

LIDAN LIU
Designaffairs China
Shanghai, China

MARTIN LOTTI
Global Creative Director
Olympics NIKE Inc.
Oregon, United States of America

CARL GUSTAV MAGNUSSON
CGM Design
New York, United States of America

EUNJOO MAING
KIDP (Korea Institute of Design Promotion)
Seoul, South Korea

STEPHANIE PEMBERTON
Australian International Design Awards
Sydney, Australia

PRINZ M. PINAKATT
The Coca-Cola Company
London, United Kingdom

AKE RUDOLF
International Design Center Berlin
Berlin, Germany

ROLF SCHWEIZER
Interstuhl Büromöbel GmbH
Meßstetten, Germany

JAE-JIN SHIM
Seoul Design Foundation
Seoul, South Korea

BOUDEWIJN SOETENS
VanBerlo
Eindhoven, Netherlands

JAN R. STAVIK
NORWEGIAN DESIGN COUNCIL
Oslo, Norway

JEREMY SUN
Orcadesign Consultants
Singapore

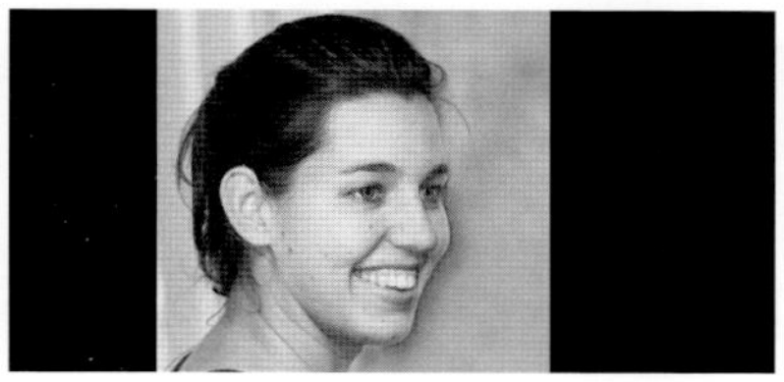

SZONJA SZESZTAI
Hungarian Design Council
Budapest, Hungary

WILLIAM TO
Hong Kong Design Centre
Hong Kong

Prof. Martin Topel
squareone GmbH
Hamburg, Germany

ADRIENNE VILJOEN
Icsid regional advisor, Sunnyside
Pretoria, South Africa

SIMONE DE WAART
Material Sense
Eindhoven, Netherlands

JURY
iF CONCEPT DESIGN AWARD 2012

PHILIPPE GROHE
DR.-ING. WINFRIED HEUSLER
KYUNG-HOON LEE
ULI MIX
GERALD MUNDERLOH
ACHIM NAGEL
JUM NAKAO
SUNGHEE PARK
ANDREAS ROTZLER

PHILIPPE GROHE

Philippe Grohe (*1967) completed an apprenticeship as a photographer and then went on to study Business Management at the "Berufsakademie Villingen-Schwenningen". During this time, he was also involved in photography, among other things, for humanitarian organizations and many different exhibitions and shows. After finishing his studies in 1995, he joined Hansgrohe USA. After working in different capacities at Hansgrohe GmbH & Co. KG and Hansgrohe AG, he has headed Axor, the design brand of the company, since 2001.

Statement

"This year, I was a judge for the second time and was once again extraordinarily impressed by the number and quality of the entries – by the enormous creative potential there is here. I saw exciting, environmentally compatible ideas and solutions in the category of Architecture. In terms of the products, however, there were a number of entries that I had seen before. It is also really quite a big challenge to evaluate minority topics in particular in the right light. This day of judging proved to me once again how important it is to actively support young international design talent."

Dr.-Ing. Winfried Heusler (*1955) has been Senior Vice President Engineering at Schüco International KG in Bielefeld, Germany, since 1998. He studied mechanical engineering at the TU Munich and earned his doctor's degree at the TU Berlin in the field of "natural daylight". From 1981 to 1998, he was employed as a development engineer and department head and then as head of aluminum siding at GARTNER, a siding company in Gundelfingen, Germany. He has become very well known among the experts due to the many lectures he has given worldwide on the topic of facades.

Statement

"My expectations were exceeded by far! I am surprised what depth some of these concepts exhibited, how intensively the participants had worked on their ideas. I saw a relatively large number of solutions that would be fairly easy to take from the drawing board to the production hall without a lot of time, trouble and expense involved. The zipper with the button, for example, is a really great concept for the fashion industry. This idea has been well thought-through and could be put on the market quickly and easily. The entries coming from Asia tend more towards multifunctionality. Is that a trend that has already come and gone or is that what we see for the future?"

DR.-ING. WINFRIED HEUSLER

Kyung-Hoon Lee heads the subsidiary of Samsung Design in Milan, Italy, since 2010. From 2012 he is also in charge of Samsung Design Europe in London. After joining Samsung Electronics Co., Ltd. in Seoul, Korea, in 1994 he developed numerous market-leading products and international prize-winning concept design products. Before moving to Italy in 2012 he was the Design Group Leader developing B2B business and network-oriented products. He earned his Master's degree in Industrial Design from Seoul National University in Korea.

Statement

"I'm very happy to be back here again and to see the enormous creativity displayed by the young design talent. This year, there were a lot of entries from the IT sector – whereby the ecological aspect was also often an integral part of the concept. There were also a lot of submissions from the medical sector. I am very impressed by this development – to see that young design talent is trying to solve real problems from a design point of view. However, design in this sector still requires considerable improvement."

KYUNG-HOON LEE

ULI MIX

Uli Mix (*1968) has been in charge of the Design Strategy division in the Volkswagen Design Center Potsdam since 2005. His responsibilities include strategic design trends for vehicle projects and concept development up until market introduction. After studying Product Design at the Karlsruhe University of Arts and Design and at the Berlin University of the Arts, he began his professional career at Audi in Ingolstadt in the Interior Design department in 1998. He is also involved in different teaching engagements in the field of design.

Statement
"iF's high international profile becomes clearly apparent when we look at the large number of Asian entries. It's interesting to see how important it is to the Asian students to measure themselves on an international scale. All in all, the quality of the entries was very high; I saw a lot of very original ideas. Some fascinating concepts were submitted once again this year; they concentrated on such current topics as protection in the event of natural disasters or providing food and shelter to refugees. The number of entries dealing with impaired vision was also very impressive – really innovative, surprising and detailed concepts."

GERALD MUNDERLOH

Gerald Munderloh (*1964) has worked for CeBIT in Hannover since 1999 – belonging to the Management Board of this major tradeshow since 2004. After studying to become a communications management expert, he, who originally trained as a wholesale and export merchant and an insurance broker, joined the advertising industry in 1991. First at Mecklenburgischen Versicherung, then at AKS Werbegesellschaft, both located in Hannover, before he joined Deutsche Messe AG in 1999.

Statement

"I was really able to sense the topic of 'design' and experience it as a link, a living element. The jury day today fascinated me. Some of the ideas, their design and their technical presentation were very impressive. This competition has shown me how much creativity there is out there in the world. From the point of view of the consumers, some of the product ideas were very easy to understand and imagine as actual products."

Achim Nagel (*1959) studied architecture at Hannover Technical University. After his activities at the Schweger und Partner architectural office, he was the manager of the central construction department of Bertelsmann AG from 1988 to 1993. From 1993 to 1999, he was a partner of the Ingenhoven Overdiek + Partner architectural office in Düsseldorf before founding PRIMUS Immobilien AG and PRIMUS developments GmbH there in the years 2000 and 2001, respectively. This was followed by the founding of the TWO product development and integrated design agency in conjunction with Claudia de Bruyn. In the year 2005, he founded möser projekt GbR with Jürgen Möser which was followed by the founding of ACP Planungsgesellschaft mbH & Co. KG with Christian F. Heine in 2009.

Statement

"Many of the entries were devoted to the topic of "responsibility" – for the environment and nature, for the young and the old living together or for the homeless. Most of these are solutions that are not technology-driven – or if technology does play a role, then it is used in people's best interests. I have seen a lot of refreshing and new, spontaneous and amusing, simple and unconventional, obvious and effective ideas. The usefulness of the concepts becomes quickly apparent – few details, simply good solutions."

ACHIM NAGEL

Jum Nakao is a fashion designer and Creative Director. He lives and works in Sao Paulo and is among the most well-known fashion designers in Brazil. His works are part of the most important fashion shows worldwide. They are on display at the Museum of Fashion in Paris and at many other museums throughout the world. He is also famous for his performance featured at the Sao Paolo Fashion Week 2004. It ended with the models tearing paper clothes from their bodies – destroying more than 700 hours of work in just a few seconds.

Statement

"I saw many interesting projects that had to do with the topics of 'rescue', 'survival' or 'medical care'. Many important design solutions that I don't see in Brazil. It was really exciting to see how the students work and express their feeling of responsibility. I was particularly impressed by the human aspect as an indispensable approach to producing good design."

JUM NAKAO

SUNGHEE PARK

Sunghee Park works at LG Hausys design center as Vice President. She studied visual communication design at the Hong-ik University in Seoul, Korea and earned a Master's degree from the Pratt Institute in New York. She worked in the cosmetic product design field at Amore Pacific, where she designed and developed many products.

Statement

"The iF concept design award has become one of the world's most important student competitions for young design talent – particularly for students from Asia. It provides a perfect platform to set apart their work from the crowd. We saw some excellent design. Great ideas, some of which were very impressively presented and explained. The students often focus on topics relevant to social and environmental issues. I saw a lot of entries designed specifically to meet the needs of the visually impaired or people with health problems."

ANDREAS ROTZLER

Andreas Rotzler graduated from his design studies in Basel with a "summa cum laude" degree. From 1992 to 1994, he worked as an illustrator for various companies and agencies, such as Leib+Gut in Bern, Rucker Huggins in California, USA and Designalltag in Zurich, Switzerland. He joined Interbrand Zintzmeyer&Lux in 1996 and is today the Chief Creative Director. Some of this company's most important clients include: Deutsche Telekom, Schindler, MINI, BMW, DOW XLA, Lucerne Festival, Graubünden, Franke, Jura, Roche.

Statement

"It was very clear to see that many of the entries deal with disasters and their effects – reactions to tsunamis, nuclear disasters or earthquakes. An incredible number of new, good approaches that I have never seen before. A bed, for example, that closes up to automatically create a protected space when it detects the vibrations of an earthquake to protect the person sleeping in it – what a fabulous idea! I noticed that some of the product designers invent 'things', when the better and easier solution would have been an app. They have to learn to think more about how to integrate their ideas in existing technology."

BECOME GOLD SPONSOR FOR THE iF CONCEPT DESIGN AWARD! WHY?

"Samsung Electronics believes that an enterprise's most vital assets lie in its design and other creative capacities. With this belief, Samsung has been cultivating the new design talents. Training and recruiting the young design talents plays a key role in enhancing the corporate competitiveness, as well as national.

We hope that the iF concept design award, an internationally recognized design competition, will provide a platform for the young designers to be acknowledged of their professionalism and to grow further to the next level, with wider perspectives and insights."

DONGHOON CHANG
HEAD OF DESIGN STRATEGY TEAM
SAMSUNG ELECTRONICS

"Volkswagen is a company with a long tradition in which design and innovations have been among the most important factors securing the company's competitive edge. Thanks to the democratization of design, innovations are now being made available to the masses. Design always has to move a step forward, it always has to be fresh.

This is one of the major reasons why we support young talent. And really new things can only be developed on the basis of an interdisciplinary, open design competition – like the iF concept design award. This award for young talent is an important instrument – for young designers as well as for industry. That's why we have already been supporting one of the leading institutions for the promotion of international design culture for many years."

KLAUS BISCHOFF
HEAD OF
VOLKSWAGEN DESIGN CENTER WOLFSBURG

"This is a great opportunity to introduce LG Hausys to young and talented designers worldwide.

The interaction of young designers with LG Hausys will allow them to gain more knowledge of Asian companies as well as Asian culture, and perhaps this intertwinement can trigger a larger picture. In addition, this sponsoring project is an excellent chance for LG Hausys to understand the current trend as well as the concerns of these young designers.

Visionary designs sprout from innovative ideas that are eventually made physical in the real world to create a better world; therefore, supporting these gifted designers and giving them a chance to display their design will help benefit not just designers but also the future of human society."

SUNGHEE PARK
LG HAUSYS DESIGN PARK

SAMSUNG

LG Hausys

We have been promoting young designers for years, which resulted in founding the iF DESIGN TALENTS GmbH. A non profit organization exclusively dedicated to young creatives. There would be a limit to our possibilities, but fortunately there are continuous and growing numbers of supporters. Why are our GOLD Sponsors convinced that this is a great, an important, a feasible undertaking? They let you know their reasons here.

"Here at Axor and Hansgrohe, we benefit a great deal from the ideas produced by our designers – and not only the established ones, the designers with whom we develop our bathroom collections. It also applies to junior designers, with whom we collaborate in a range of different projects. The wealth of creative potential coming to the fore was once again evident in the many terrific works and ideas that the members of the judging panel for the iF concept design award were asked to review and assess. Being a member of this panel has again been a lot of fun for me. On each occasion I learned a great deal, and I am very pleased that we are able to actively support and therefore contribute to the success of what is arguably the most prestigious award for junior designers from around the world."

PHILIPPE GROHE,
HEAD OF THE AXOR BRAND/HANSGROHE SE

Technology and design are inextricably linked. As the organizer of CeBIT, the world's most important event for the digital economy, promoting and encouraging young designers is very close to our heart. Modern design concepts are key drivers of innovation and success in today's high-tech industry: in addition to visual appeal, they address functionality and user friendliness. As a sponsor of the iF concept design award, we support talented young artists whose creative work brings the world of technology to life. We wish all nominees much success with their pursuits.

FRANK PÖRSCHMANN,
MEMBER OF THE BOARD RESPONSIBLE FOR CEBIT,
DEUTSCHE MESSE AG

"Young designers are the future and even more so as they help create it. The ideas and concepts entered in the iF concept design award give us a glimpse of that future – of small solutions as well as approaches in resource saving, demographic changes and environmental friendliness. That is why we have always insisted on supporting young design talents. Also, this competition is bridging a gap: economy and industry will find innovative approaches, fresh minds and even some concepts worth to be taken seriously within the scope of a critical analysis. Talented young designers can compare, compete and be seen on an international level, even come into direct contact with companies at an early point in their careers. We take promoting young talent serious."

RALPH WIEGMANN
iF MANAGING DIRECTOR

SAMSUNG

Make it Meaningful

Samsung is working in a "value focused" collaborative environment that allows them to exchange meaningful ideas. The aim is to inspire one another to create products that can deliver meaningful customer experiences, better their lives and the world that they live in.

Samsung's Smart TV is different than conventional models.
It becomes one with the space around it, beautifully designed to blend into its environment.
Samsung's D8000 uses only metal material to interpret the Narrow Bezel design.
The distinction between the TV and background appears to be only that of a single line, delivering a premium natural quality.
The combination of metal that joins the unit's main body to the stand highlights the unit's aesthetics and has never been showcased in televisions before.

»Der up! verkörpert eine Kultur, in der sich Größe nicht allein in Dimensionen oder Leistung ausdrückt.«

Walter de Silva, Head of Volkswagen Design AG

VOLKSWAGEN DESIGN

Das Auto.

Der neue up!

product
design award
2012 GOLD

Designed by Zaha Hadid Architects

Nature, Human and Space

By making customer value creation its top priority, and by pushing for relentless innovation, LG Hausys is a company that creates living spaces which enrich our customers' lives, bringing about an unprecedented balance between nature and people.

LG Hausys

As Korea's largest provider of interior building materials, LG Hausys has been producing happy and rich living spaces for more than 50 years. This has been accomplished through the use of market-leading products and advanced technology along with the use of eco-friendly and aesthetically pleasing materials.

Starting with the production of vinyl sheets in 1958, the company has contributed to enriching people's lives for more than 50 years by supplying distinguished products and services that include Hi-Sash (Korea's the first plastic windows), High Gloss sheets (which occupy the world's biggest market share), HI-MACS® (acrylic solid surface), and building-integrated photovoltaics (BIPV).

By applying the industry's leading technologies, LG Hausys designs spaces that are as beautiful as they are functional, using eco-friendly and beneficial products that command a dominant market position in their respective fields. Through successful localization strategies that have seen outstanding performances in the U.S., China and Russia, LG Hausys is constantly solidifying its stance as a leading global company.

In addition, the company is carefully preparing itself for the future through continuous R&D and increased investments in facilities, as well as through strengthened marketing & design capabilities.

LG Hausys Design Center

Knowing that design will play an increasingly important role in corporate competitiveness, LG Hausys Design Center is emphasizing creativity to produce innovative designs.

Keeping in line with our vision of producing pleasing living spaces that are natural and filled with humanity, LG Hausys Design Center strives to enrich the living spaces of our customers through the evolution of design. In this way, our products possess not only superior design, but also an emphasis on life and cultural merit. With our devotion to customers and our insight of the future, LG Hausys Design Center will create more pleasing living spaces.

LG Hausys

Hansgrohe SE Making life in and around

he bathroom better and more beautiful

For more than a hundred years, we at Hansgrohe, mixer and shower specialists from the Black Forest, have been working with water. Some forty years ago, along with Hartmut Esslinger, we began introducing good design to the bathroom. Back then the bathroom was a functional sanitary cubicle. To most people, the notion of gladly spending more time than necessary there, out of a desire to feel good, was still very strange.

Our classical products from those days, such as the colored Tribel showers, are still in demand. Since those times, the company has continued to develop innovations and set strong trends. For the past twenty years, along with our designer brand Axor, we have developed visions, concepts, and exclusive solutions for the bathroom as a living area.

What unites our two brands, Hansgrohe and Axor, is the wedding of excellent design, technological innovation and uncompromising quality for the benefit of our customers. We work only with outside designers – based on the conviction that their experience, also from other industries, will repeatedly afford new perspectives, and will also keep creativity alive for a long term. So you won't find an in-house design department at Hansgrohe SE.

Aside from this shared understanding, the profiles and design philosophies of the two brands are quite distinct: whereas Hansgrohe, the premium manufacturer, covers the middle and higher-priced segments, Axor is positioned in the top market segment. While Hansgrohe has been successfully developing mixers and showers together with Phoenix Design, Axor has cultivated its own, more holistic approach together with strong designer personalities.

Our focus is on the bathroom as a living area and the people who are meant to feel comfortable there. How should they use and sensually experience water? Which emotional needs do they associate with the time spent in the bathroom, and what are their functional demands? Based on this premise, outstanding product designers, but also architects and interior designers, create for Axor visions of their future bathroom. In an open, creative dialogue, we jointly develop trendsetting collections, along with space and utilization concepts – solutions to contribute towards making life in and around the bathroom that little more better and beautiful.

hansgrohe

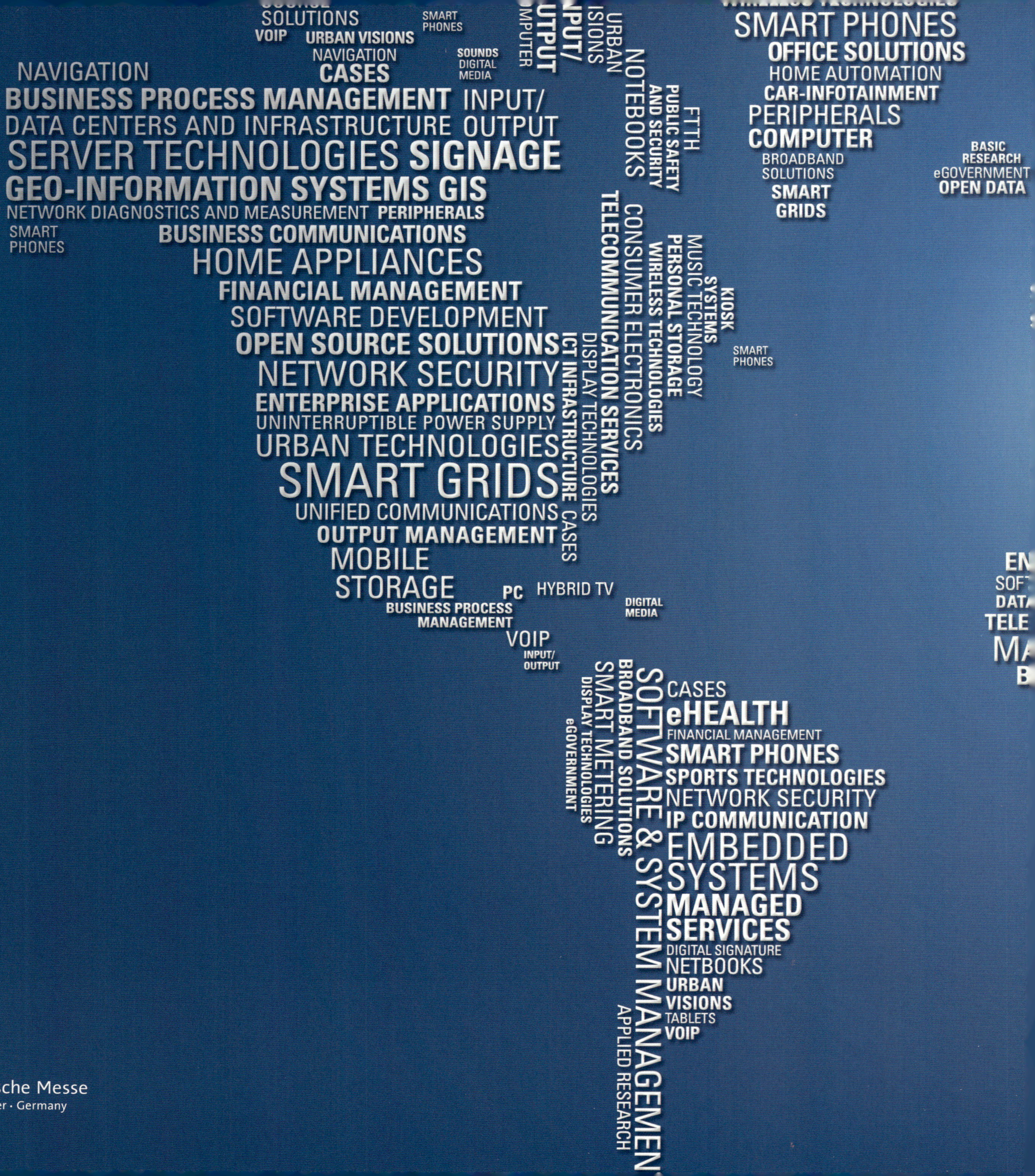

Deutsche Messe
Hannover · Germany

PUSH
YOUR BUSINESS
VIRTUALIZATION
OUTSOURCING
WEB 3.0 - INTERNET OF THINGS & SERVICES
NETWORK DIAGNOSTICS AND MEASUREMENT
SOFTWARE & SYSTEM MANAGEMENT
UNIFIED COMMUNICATIONS
DIGITAL SIGNATURE MANAGED PRINT SERVICES
URBAN TECHNOLOGIES MOTHERBOARDS
CAR-INFOTAINMENT PERSONAL STORAGE CUSTOMER RELATIONSHIP MANAGEMENT
ICT SERVICES eGOVERNMENT
KIOSK SYSTEMS URBAN VISIONS
ENTERPRISE APPLICATIONS
eHEALTH CLOUD COMPUTING
COLLABORATION CONSUMER ELECTRONICS
HOME APPLIANCES OFFSHORING
SPORTS TECHNOLOGIES NETWORK SECURITY
BUSINESS PROCESS MANAGEMENT SOFTWARE DEVELOPMENT
SOUNDS UNINTERRUPTIBLE POWER SUPPLY
MOBILE SOLUTIONS
AMBIENT INTELLIGENCE
BUSINESS COMMUNICATIONS
CeBIT
cebit.com
HANNOVER
5.–9. MÄRZ 2013
Heart of the digital world

1953 – 2013

60 YEARS iF DESIGN AWARDS

PURE EXCELLENCE

iFDESIGN.DE

AWARD WINNERS, PRIZE MONEY

44 international professionals looked at every single one of the 10,665 valid entries from 62 countries to finally shortlist 300. Another nine jurors were then invited to examine and discuss this shortlist. They choose the "Best 100" – the 100 entries to win the iF concept design award 2012. The jury then selected the 20 most outstanding submissions and divided the prize money, a total of EUR 30,000. You will see how on the next pages: starting with the highest prize money winners. For the first time, jurors often left comments as to why they had chosen the entry, visible in grey letters. The winners' reaction can also be seen at the bottom of each winning entries page.

SAMSUNG
LG Hausys
A PROJECT OF
DESIGN TALENTS

industrial design + product design

Design
Daniel Nikol
Bamberg, Germany

Hochschule Coburg
Coburg, Germany

Prize money EUR 5,100

:NTER:M
DISASTER RELIEF HOUSING

Natural catastrophes usually hit the poorest of the poor the hardest. ":inter:m" is a folding emergency shelter that can protect up to five people from harsh weather conditions. The concept focuses on simple assembly and transport as well as affordably priced production in larger numbers.

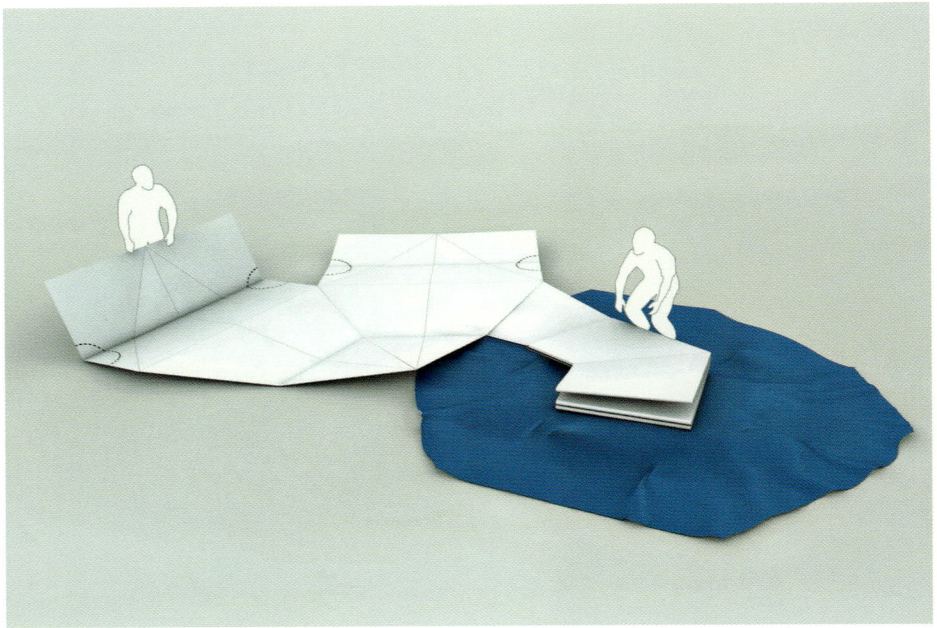

Achim Nagel: *"This concept proposes using recycled materials to make a folding shelter that is extremely practical in terms of use and transport. Perfect to meet the needs of the homeless or victims of natural disasters. A particularly convincing concept. Congratulations!"*

Boudewjin Soetens: *"Well presented and an idea for a good cause. Will need some detailing, but works as a concept."*

"Seeing how people have to improvise in disaster zones as part of my concept was the most exciting finding during the work on :nter:m."

industrial design + product design

Design
Rebecca Loaíza
Kiel, Germany

Muthesius Kunsthochschule
Kiel, Germany

Prize money EUR 4,100

CONFIO
MEDICAL DEVICE: DIAGNOSIS OF THE GI TRACT

"CONFIO" is a diagnostics system for the digestive tract. The patient swallows the CONFIO capsule, which contains several miniature cameras. The capsule has a wireless connection to a control unit to provide the physician with a three-dimensional view of the small intestine in real time. The patient requires no sedation and doesn't need to undergo an unpleasant procedure.

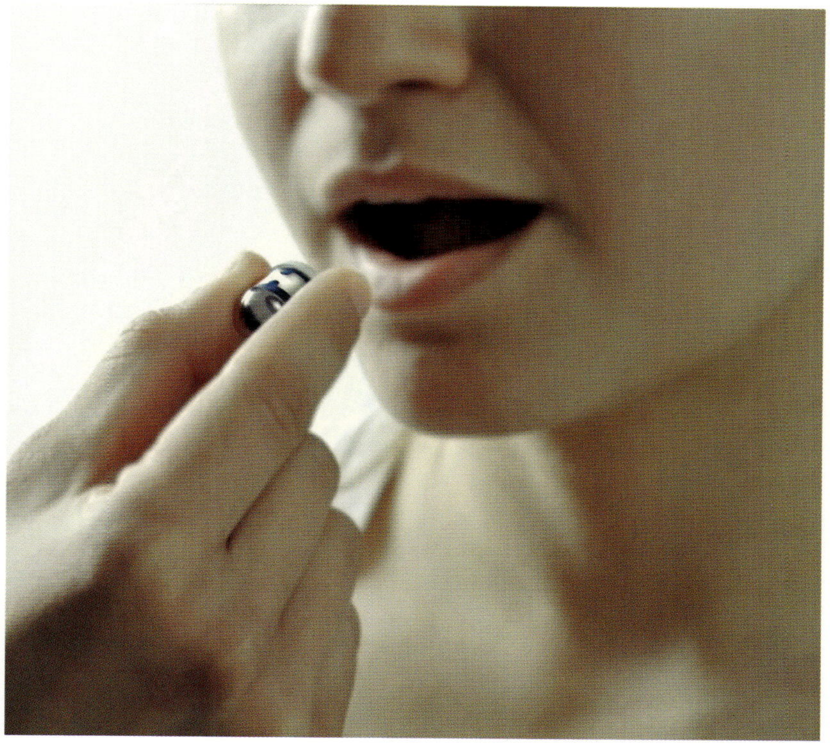

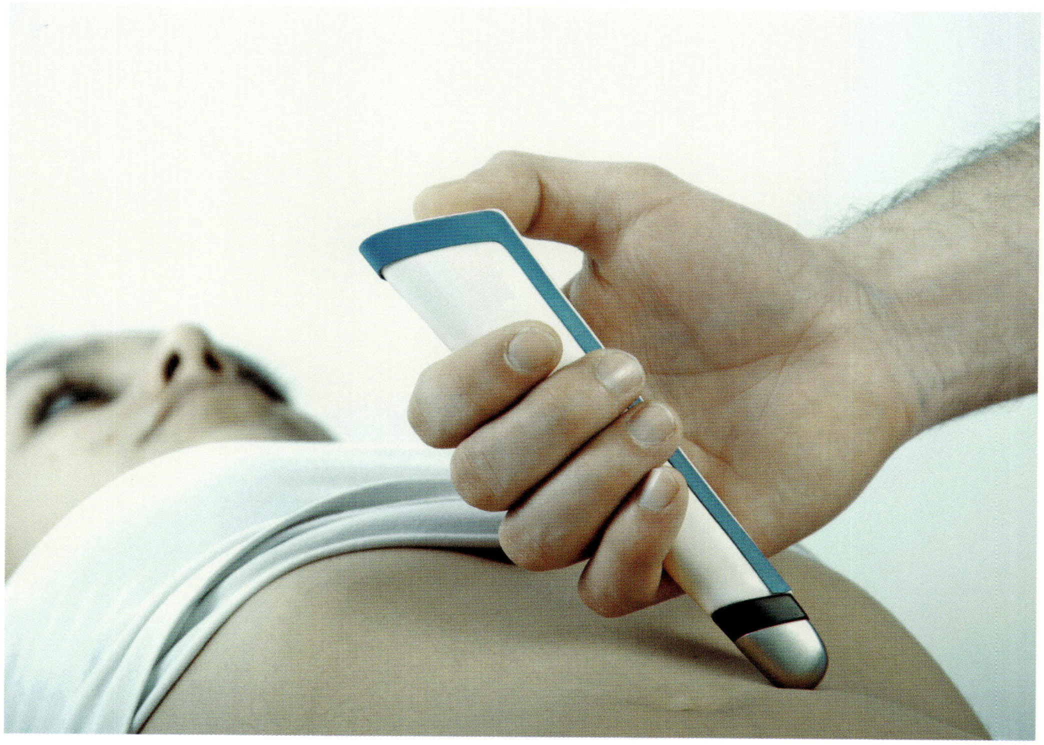

Uli Mix: *"This special, non-invasive concept saves the patient a lot of unpleasantness. It is no longer necessary to swallow a hose. An interesting approach – whether or not it can actually be put into practice deserves careful consideration."*

"The appreciation shown for my work, particularly among so many other entries, has given me a huge boost in terms of personal motivation."

interior design + architecture

Design
Chih Hsuan Li
Taipei, Taiwan

National Taiwan University of Science and Technology
Taipei, Taiwan

Prize money EUR 4,100

DUCKWEED PAVEMENT
WATER PURIFYING WALKWAY

DUCKWEED PAVEMENT is made up of floating plateaus that help to purify contaminated or polluted water reservoirs. This layered system consists of individual panels that can be assembled flexibly, making them suitable for use on the surface of any bodies of water.

Achim Nagel: *"An intelligent, multi-layer system that not only purifies water, but also creates high-quality, micro-climatic green areas at the same time. A perfect solution for a serious, complex problem."*

Ake Rudolf: *"The concept proposes a convincing combination of social space, ecological habitat and sustainable functionality. Even though feasibility, expense and effectiveness still have to be investigated, this looks like a very viable concept. Could also be used to grow food or as social gardens and cultivable areas. All in all, a positive concept with great social, environmental and developmental potential."*

"I can't imagine any words to describe what I am feeling right now. I deeply appreciate iF juries giving this chance to me. I will keep studying more about sustainable environmental strategies. And I will never forget as a designer that we can create a better future for mankind."

industrial design + product design

Design
Honghai Yu
Zhengzhou, China

Zhengzhou University of Light Industry
Zhengzhou, China

Prize money EUR 3,100

"BALANCE" SEESAW
SEESAW

Children of different weights cannot play together properly on conventional playground seesaws. "Balance" solves this problem by using sides of variable lengths. The lighter child simply pulls his/her seat further out – and the weight is "automatically" balanced out.

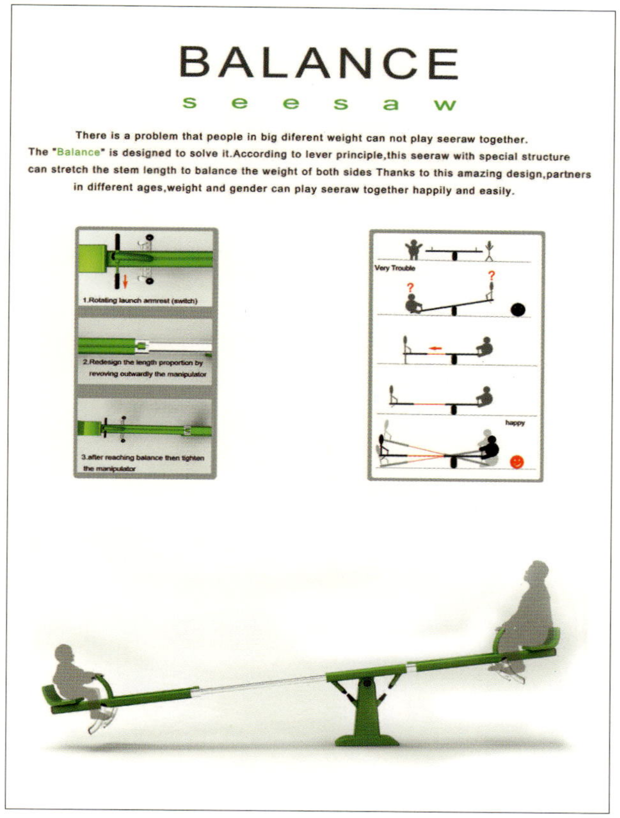

Arhan Kayar: *"It's good to show and teach the balance system."*

Gerald Munderloh: *"A neat solution that allows children of different weights to play together. Particularly valuable from a social point of view because size and weight are not important. A perfect combination of design and functionality."*

„Winning the iF concept design award 2012 is a positive response to our work as well as an honor for us. Thanks to our instructor, Chien-Ming Hsieh. Under his guidance, we were able to think of so many creative ideas and experienced good team spirit. Thanks also to our university and department for providing a platform of design for us to learn and to go beyond the normal limits."

industrial design + product design

Design
Matej Colja
Komen, Slovenia

University of Ljubljana
Ljubljana, Slovenia

Prize money EUR 1,300

GARDENWALL
GARDEN RACK

GardenWall is a greening system for urban areas, with which vegetables can be grown throughout the entire year. This element can be mounted on any façade or used as a partition in an apartment. A special photovoltaic glass produces electricity and shade if the sun gets too hot. Temperature, oxygenation, watering and lighting can be controlled wirelessly via cell phone.

Achim Nagel: *"This extremely useful façade greening system could be used for new as well as for existing buildings. The inhabitants could produce some of their food themselves and positively influence their micro-climate at the same time. A great idea for nature enthusiasts who live in the city."*

Ake Rudolf: *"A very doable product in different versions with a number of different positive features – from growing food to generating power, air purification, control of the micro-climate and right on up to social functionality."*

"For me personally, this award represents a significant achievement and I am very proud of it. It confirms my design thinking, my view of the future and greatly encourages me to stay in the field of design. Thanks to the entire organization and the judges who awarded me this prize."

industrial design + product design

Design
Di Fang, Xuefei Liu, Long Lu, Jinghua Guo
Dalian, China

Dalian Nationalities University
Dalian, China

Prize money EUR 1,300

MAGICAL BUTTONS
BUTTONS

If you forget to zip up the fly of your pants, things can get embarrassing. That will never happen again with the "Magical Button". The zipper is also the pants button.

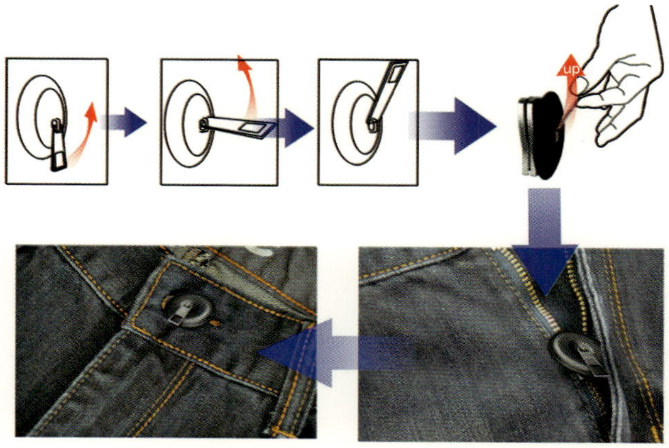

Dr. Winfried Heusler: *"A great invention that makes sure the fly stays closed. A small, but clever idea that can develop into a real USP in the fashion industry."*

"Thanks to the jury on the recognition of our design, we will continue to work hard for people to design better things."

industrial design + product design

Design
Changjun Park, Byungeui Lee, Munhan Bae, Jungsu Kim
Seoul, South Korea

Seoul National University of Science and Technology
Seoul, South Korea

Prize money EUR 1,300

MEDICAL KIT
MEDICAL PACKAGING

There is a critical lack of hospitals and medical care in areas where natural disasters have occurred, in crisis regions and in poor parts of the world in general. "Medical kit" is a cardboard display set-up, which can be used to give the injured and the ill intravenous injections quickly anywhere at all. This is a quick and effective way to save lives. After use, the display can be simply folded up again.

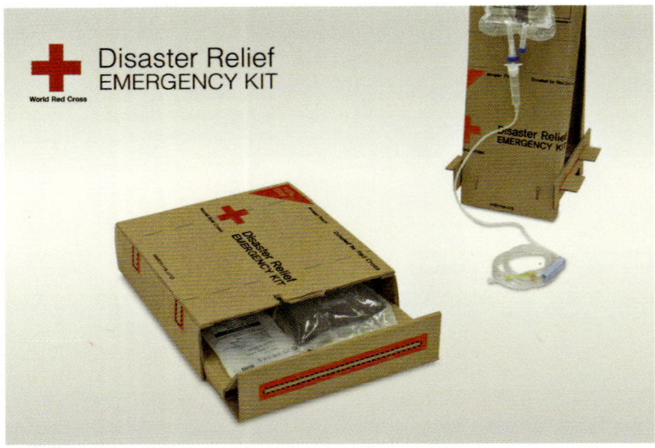

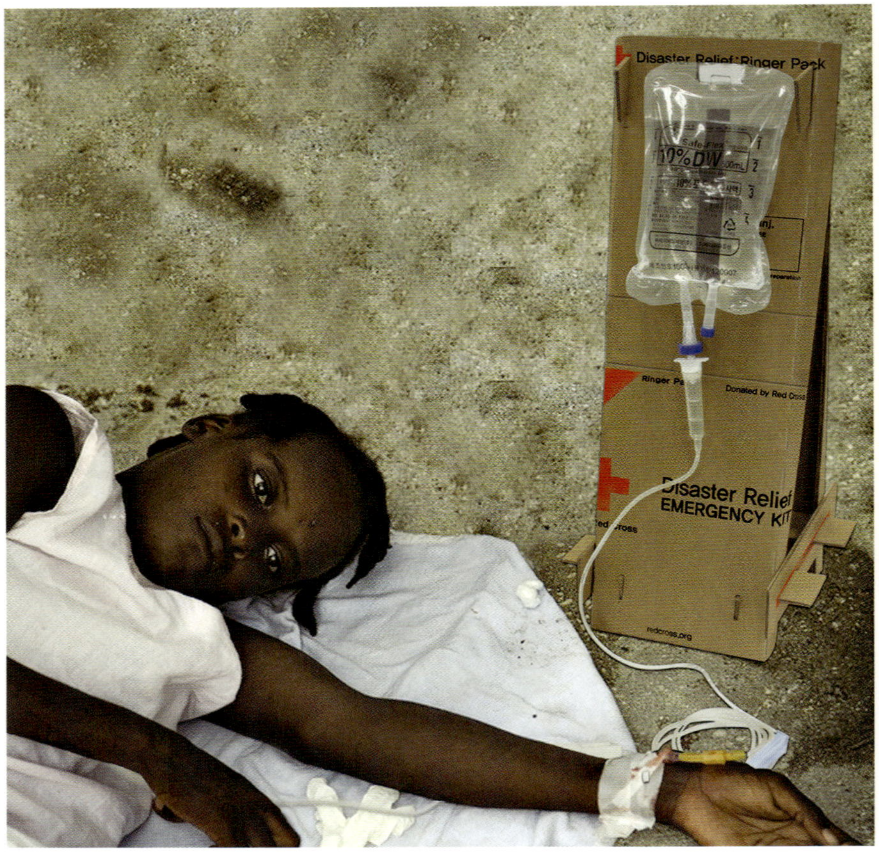

Sunghee Park: *"A simple medical idea that can save lives. This kit, which can be used over and over again, is an excellent solution for situations that require immediate medical attention."*

"I feel quite surprised. It's an honor to win this award and I really wanted it.
I am going to think about "Share-Design" which I dream through this award.
Wish that I will get a great potential for growth from this opportunity. THANK YOU!"

industrial design + product design

Design
HeeWon Kim, SoHee Han, JiYoung Lee
Gyeong-san, South Korea

Catholic University of Daegu
Gyeong-san, South Korea

Prize money EUR 1,300

PULL & INVERT UMBRELLA

To close this umbrella, the tip of the umbrella is pulled towards the handle. The remaining rain water is collected in the closed umbrella and the user is not sprayed with this water in the process. This umbrella design also makes it easier to get into a car while it's raining – without getting the umbrella caught in the door.

Jum Nakao: *"A very unusual, clever solution for an everyday product! It clearly proves how important new, innovative design is for industry."*

Peter Kuchnicki: *"Wow – why wasn't this invented 100 years ago! Smart and practical, this innovative design manages to solve several problems. Gyeong-san can be proud of their young designers."*

"We are really excited that this product concept won the iF concept design award 2012. Our original intention was to learn more about product design through the competition and we didn't expect to win. We will continue to gather experience and study more about design and how it can be used to make life better."

industrial design + product design

Design
Byunghoon Ju
Yongin, South Korea

Dankook University
Yongin, South Korea

Prize money EUR 1,300

RABBIT OUTLET
ELECTRIC OUTLET

Power outlets are frequently installed so low on the wall that it is difficult to see where the plug fits in. The "rabbit outlet" with the two ridges on the casing is designed so that the plug can be put directly into the outlet – from any direction.

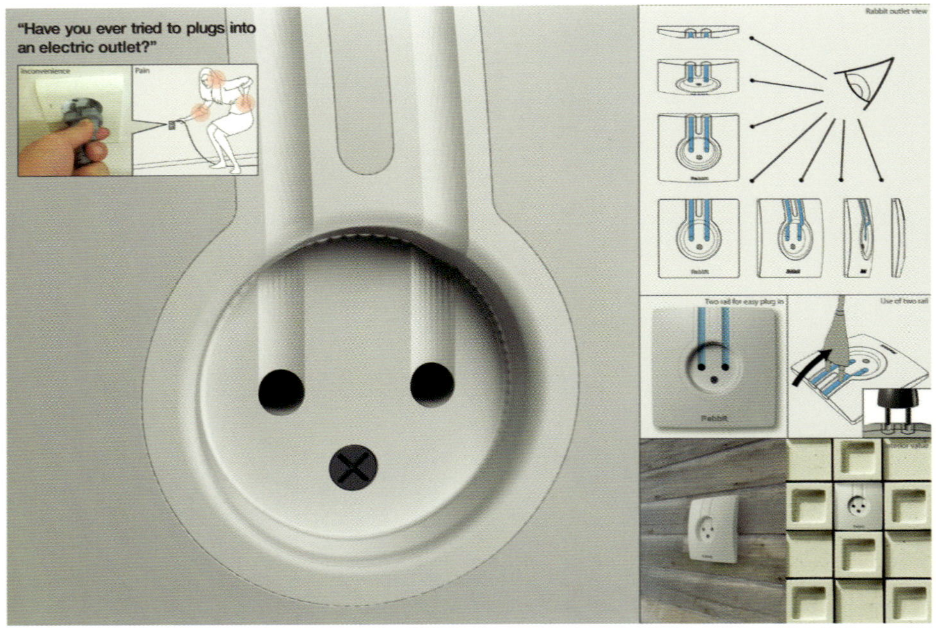

Gerald Munderloh: *"This idea for the design of a power outlet is a simple, but effective solution – a perfect combination of design and functionality."*

Tiago Krusse: *"Simple and intuitive product; functional and visually appealing."*

"I am so happy that my design was recognized in this honored award.
I have a feeling my career's just begun. Thank you."

industrial design + product design

Design
Da-Hee Jeong, Chi Gook Lee
Gunsan, South Korea

Hongik University
Jochiwon, South Korea

Prize money EUR 1,300

SADDLE LOCK
BICYCLE LOCK

Bicycle seats are often stolen because they are so easy to remove. "Saddle lock" can solve this problem: the seat post has an extension that is flexible. This makes it possible to use the saddle with the seat post as a bicycle lock.

Andreas Rotzler: *"A clever idea that solves a simple, everyday problem. The concept still needs a bit of work, but it would certainly make life easier for cyclists."*

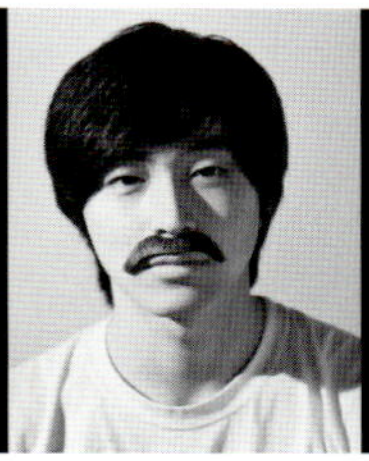

„Thank you for giving us this great award. We are really happy to receive this prize among so many competitors. Winning this prize made our relationship even better and made us more confident as designers. Again, thank you!"

industrial design + product design

Design
Guang Xi, Na Zhou, Xu Liu
Wuhan, China

Huazhong University of Science & Technology
Wuhan, China

Prize money EUR 1,300

TURN-SAFETY BED

"Turn-safety" is a special bed that can save lives in earthquake-prone regions of the world. If the bed frame detects a certain vibration intensity, two side pieces shoot up to create a stable hollow space. Should the person be buried under the rubble, this hollow space provides protection until the rescue team can get there.

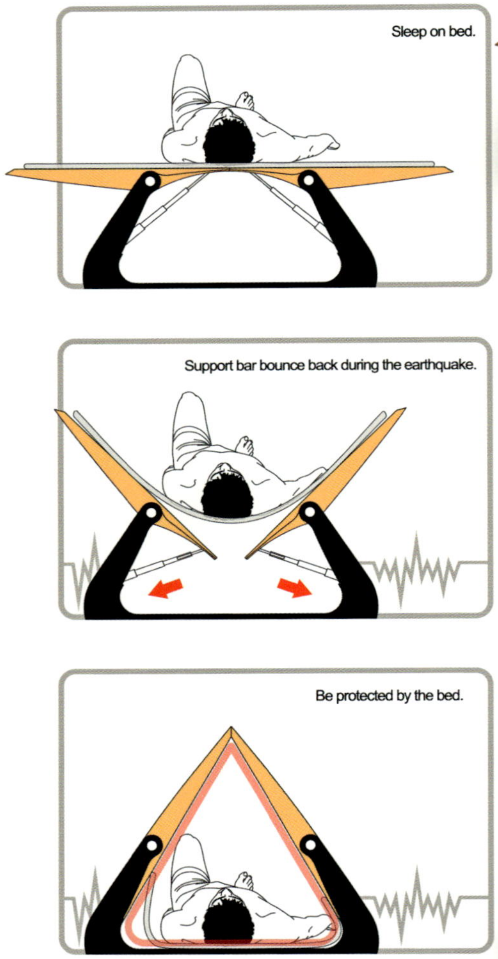

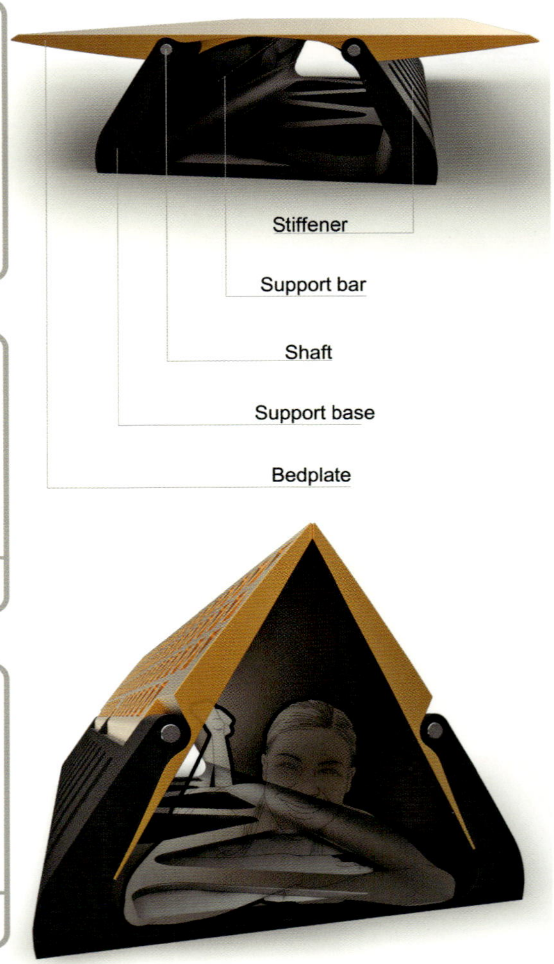

Andreas Rotzler: *"Earthquakes are particularly dangerous when they surprise people in the night. This system uses a simple functionality to quickly create a safety zone where victims can wait until help arrives. Great solution."*

"We are honored to receive this award. Disasters like earthquakes are inevitable, but we are doing our utmost to create good designs to save people's life in natural disasters. We are now encouraged to continue in this way to make design do good for humankind."

interior design + architecture

Design
Jiaxi Wang, Junjie Peng, Meiying Shen,
Leikai Zhang, Mina Ying
Hangzhou, China

Zhejiang University
Hangzhou, China

Prize money EUR 500

AD-TRAMP
PUBLIC CARE FACILITY

"Ad-Tramp" is a city-light billboard with an integrated sleeping shelter, which can be used by homeless people looking for somewhere to sleep. The illuminated billboard emits warmth that is directed into the shelter area to create a comfortable atmosphere. A solution that uses already available energy for social purposes.

Philippe Grohe: *"Not a good solution in terms of design, but a good approach to make advertising systems useful for social purposes."*

Stephanie Pemberton: *"This concept successfully converts a purely commercial venture into a socially sustainable solution to a problem that effects every city."*

"The ultimate goal of designing is to make life better for each and every participant in our society. Hence as a design team, our designs have always come from the very concern for the vulnerable groups in our society, which have long been invisible to many eyes."

industrial design + product design

Design
Jae Yoon Park
Seoul, South Korea

Konkuk University
Chungju, South Korea

Prize money EUR 500

CLENDLE
SUBWAY HANDLE STERILIZATION

Anyone who uses the grab handles in the subway always runs the risk of contracting contagious illnesses. The "Clendle" subway grab handle minimizes the danger of infection: the handle mounting is equipped with a sterilization function – if the handle is turned, the bacteria and germs are killed.

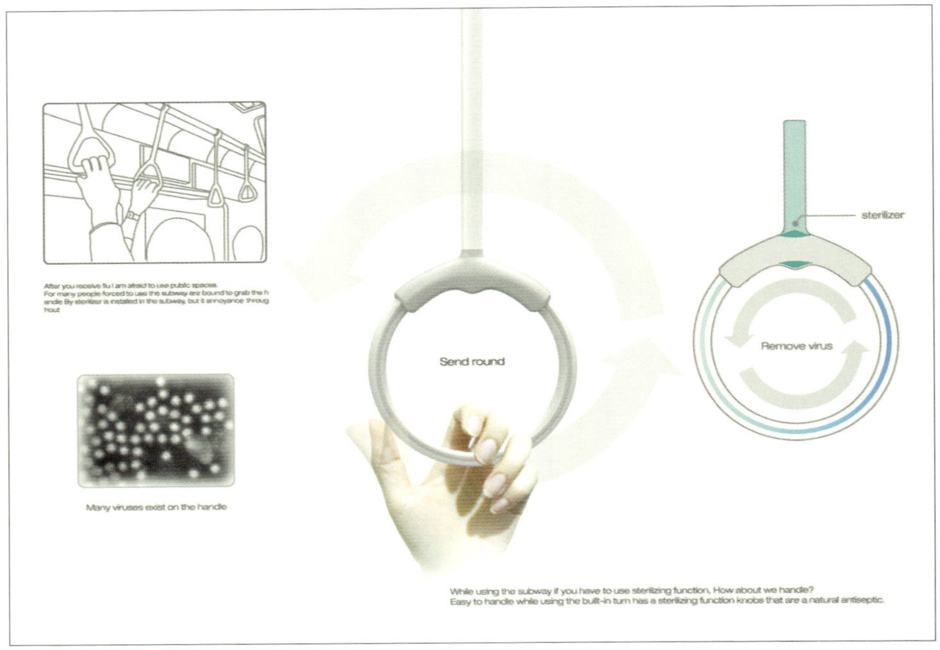

Gerald Munderloh: "Clendle" is a real benefit for the general public: the health and social aspects have been solved simply and effectively."

"The good news is wonderful and a pleasure to hear. My design work received good reviews and I am delighted. I will continue to work in this field. I want to dedicate this award to my parents, who have always believed in me. Thanks to all."

industrial design + product design

Design
Jae il Bae
Seoul, South Korea

Konkuk University
Seoul, South Korea

Prize money EUR 500

Jum Nakao: *"When cooking, it's a real advantage to get the amount of water you need without having to measure it out. This faucet is a great combination of sustainability, efficiency and practical application that makes life easier."*

Simone de Waart: *"Smart idea that reduces water waste and creates awareness for water usage; at the same time is handy during cooking. Using a knob as the best way could be reconsidered."*

COOK
FAUCET

COOK is a water-saving water faucet for the kitchen. By simply pressing a button, you get exactly the right amount of water that you need for cooking, for example – 100 ml, 500 ml or 1 liter. The longer the button is pressed, the more water comes out of the faucet.

"I am very pleased and proud to be awarded a prize in this international contest. It has been a great experience and has given me the self-confidence to enter such competitions in the future as well. Finally, I would like to express my sincere gratitude to all of you who gave me valuable advice and assistance."

industrial design + product design

Design
Woojung Moon, Wonbo Shim
Seoul, South Korea

sadi – Samsung Art and Design Institute
Seoul, South Korea

Prize money EUR 500

EASY TRIGGER
FIRE EXTINGUISHER

This very flat, space-saving fire extinguisher has been designed so that it is also very easy to use, even in emergency situations. In contrast to conventional fire extinguishers, where the locking mechanism will often not open in panic situations. To use the "Easy trigger", it is only necessary to pull out one locking pin. And the "Easy trigger" looks good, too.

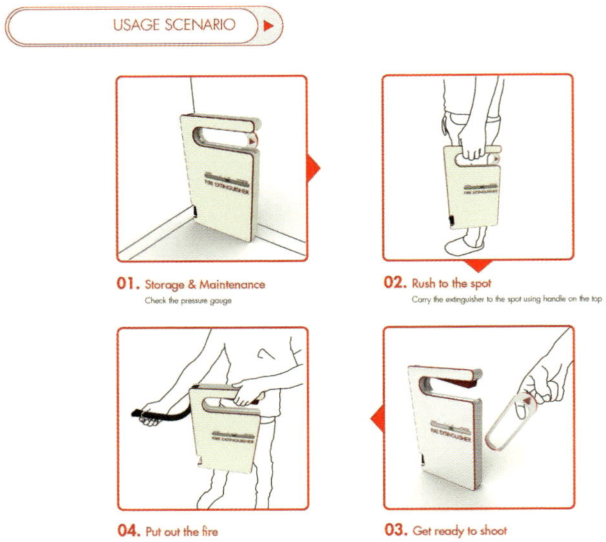

Boudewjin Soetens: *"Interesting redesign of an age-old product, both in use and in renewing the archetype."*

Dr. Winfried Heusler: *"A wonderful solution in terms of functionality and safety: the fire extinguisher is also easy to use, even in panic situations. In addition, the graphics always indicate the level to which the fire extinguisher is filled."*

"What is accepted as the norm isn't necessarily right. It's just a habit. Change people's behavior, then solve the problem. That's the approach that we take in our design. Thank you."

fashion design

Design
Uta Tischendorf
Dresden, Germany

Burg Giebichenstein Kunsthochschule
Halle/Saale, Germany

Prize money EUR 500

EVERYDAY RITUALS
FOLDED KNITTINGS

"everyday rituals" is a new technological development in the field of knitted textiles. The source of inspiration was the Japanese art of paper folding – origami. With the help of different technological knitting techniques and different material combinations, the folding knitted fabric was developed with the optimum technological parameters. Thanks to its durability, flexibility and memory behavior, it is equipped with a multifunctionally usable movement mechanism.

Jum Nakao: *"This 3D fabric is a fully new material development and a door-opener for many new processes in the fashion world. Up to now, the clothing industry has been limited to the use of conventional materials – this opens up completely new horizons for future production."*

Boudewjin Soetens: *"An interesting combination of material innovation and traditional handcraft, with a convincing result."*

"Thank you very much for the international recognition. I'm so glad about it. Bob Dylan once said people are successful if they do something between getting up and going to sleep, which he likes to do. My work is my passion, it's what I love doing. Success is not the key to happiness; happiness is the key to success. Thank you!"

industrial design + product design

Design
Chu-Hao Tso, Tsung-Hua Chuang, Hao-Tien Hsiang
Taichung, Taiwan

Asia University
Taichung, Taiwan

Prize money EUR 500

LIFE LIGHT
LIFE BUOY HOLDER

Bodies of water are secured with warning signals, life preservers or buoys. All these items are useless, however, when visibility is poor or at night. "Life light" is a combination of a buoy, a life preserver and a warning signal that is illuminated thanks to an LED water battery. In an emergency, the device can be located immediately, even at night.

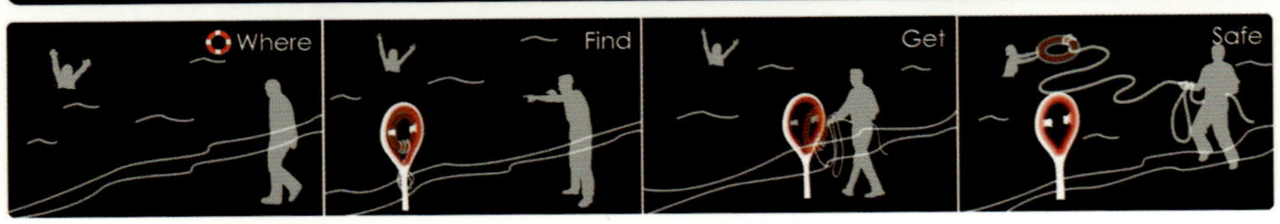

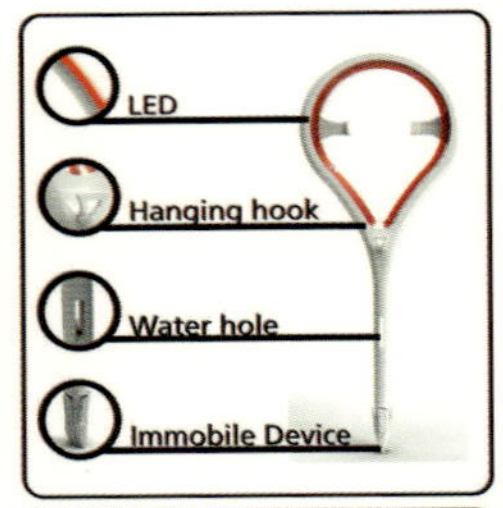

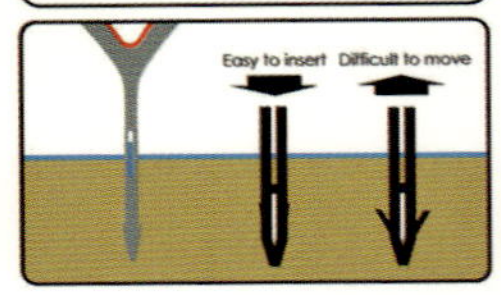

Kyung-Hoon Lee : *"Simple, intuitive operation, an idea that can save lives. Aesthetically and symbolically, an idea that needs no further explanation for its operation."*

Jan Stavik: *"A typical design for all solutions in a life-saving situation."*

"Receiving this special prize means that our creativity met with the approval of the judges. This award means a lot to us because we are getting closer to our dream of design. It has renewed our passion and encourages us to continue in the future."

industrial design + product design

Design
Inesa Malafej
Panevezys, Lithuania

Vilnius Academy of Arts
Vilnius, Lithuania

Prize money EUR 500

MY WRITING DESK
DESK

Never have an untidy desk again! "My writing desk" is a desk enclosed by a wooden construction. The wooden frame provides a lot of space to store things that usually take up space on the main desk, but still keeps them in plain view. Things that often fall off the desk also land in the wooden frame. The desk is designed so that it can be easily dismantled without tools.

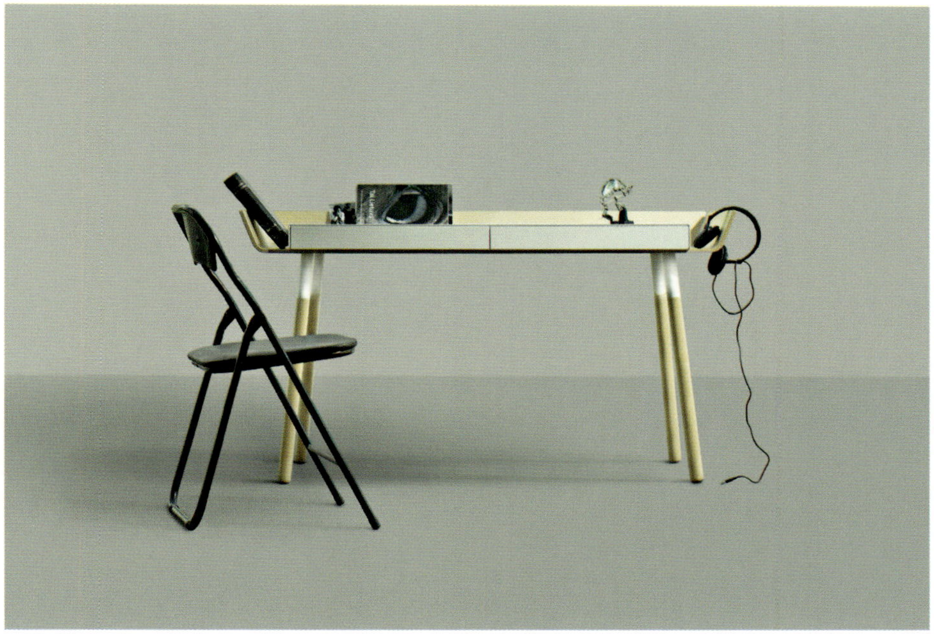

Philippe Grohe: *"An elegant solution: in addition to a larger working space, the system also offers a presentation and filing system."*

Prof. Martin Topel: *"A simple idea to keep things neat and tidy that also reflects precision of form."*

"I feel very pleased that my work is among the best designs of the year. While I am now taking my first steps in the design world, getting the iF concept design award 2012 reassures me and makes me confident that I am moving the right direction."

industrial design + product design

Design
Guang Xi, Na Zhou, Yang Shi
Wuhan, China

Huazhong University of Science & Technology
Wuhan, China

Prize money EUR 500

RHINO BIRD
DEVICE FOR BLIND PEOPLE

This intelligent, portable navigation system has been designed for the visually impaired. It guides the user to the desired destination, while warning of any dangers along the way at the same time. The small hand unit is inconspicuous and communicates the same information that could be retrieved by the sense of feeling.

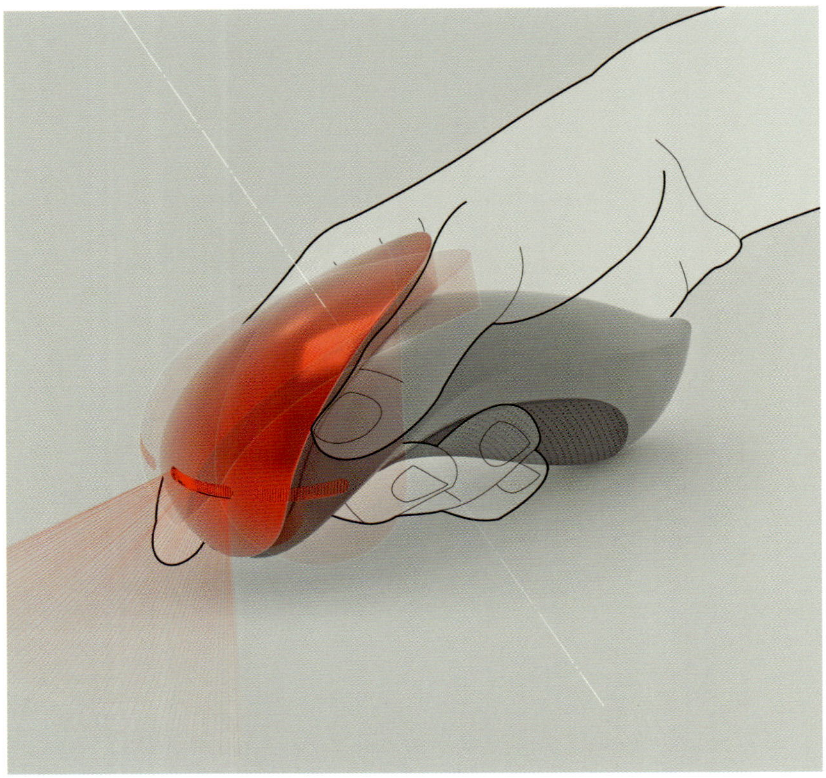

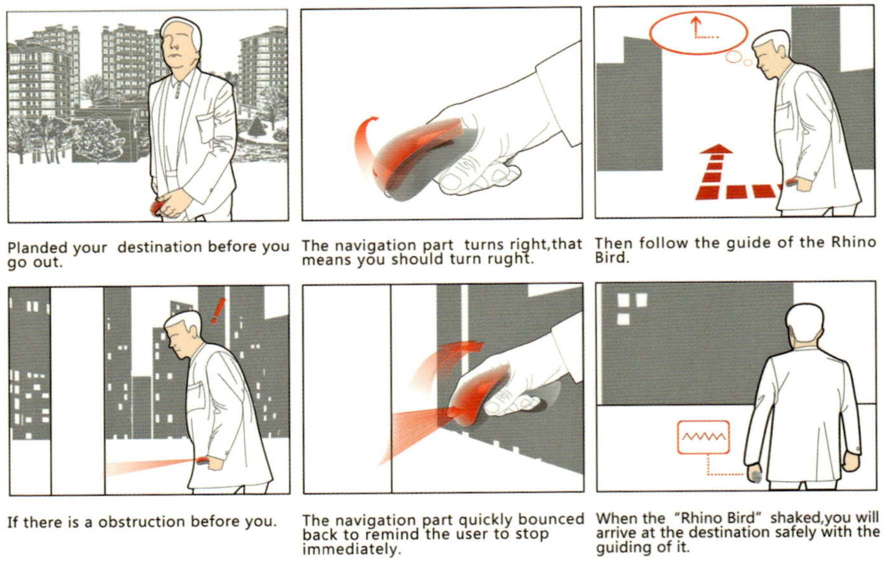

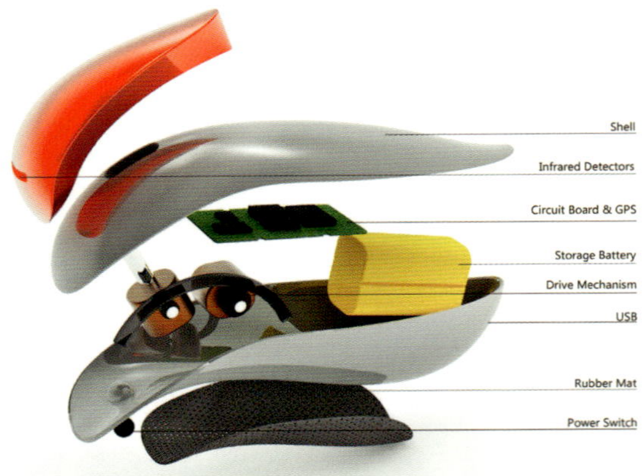

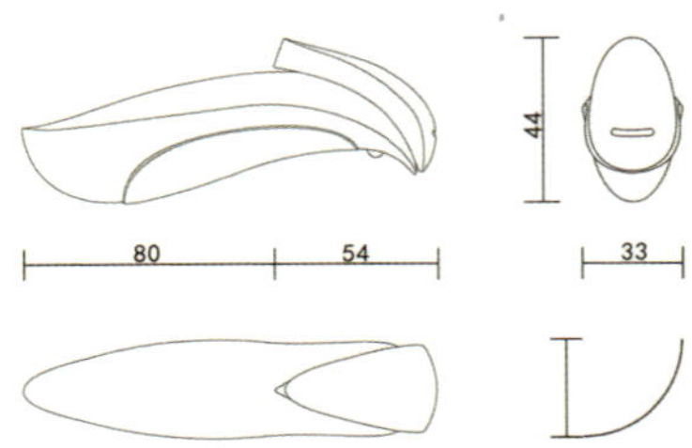

__Andreas Rotzler:__ "This combines a white cane with a seeing-eye dog. "Rhino Bird" is a handy solution for an ancient problem: orientation in the immediate surroundings and navigating from one place to another."

__Li Bei:__ "The design is warm for the blind. It adopts bionic form, GPS and infrared ray navigation technology, and the selected materials have quality and utility function. The entry has commonality and safety."

„We are extremely excited to obtain an award in the international competetion of this kind. We believe a designer's duty is to help everyone leed a better life, so our team devote ourselves to supplying products for the disadvantaged people like blind people, hoping that they can also enjoy happy life like ordinary people. This award is a confirmation of our work, which encourages us very much."

industrial design + product design

Design

David Lee, Yuna Kim (Handong University)
Daejeon, South Korea

Korea Advanced Institute of Science and Technology – KAIST
Daejeon, South Korea

Prize money EUR 500

Sunghee Park: *"Learning a foreign language is often difficult for many people – looking up individual words is very time-consuming. This product is a great idea because this is exactly the problem that it solves – quickly and easily."*

WYSIWYG DICTIONARY
DICTIONARY

Looking up a word in a foreign language while reading a book used to mean it was necessary to type that word into a computer, rather inconvenient, to say the least. With the WYSIWYG Dictionary, that is now much quicker and easier: all you have to do is to hold the small, transparent OLED screen over the text and touch on the corresponding word. Its translation appears immediately. The translation of entire sentences is also possible if you run your finger over the full sentence.

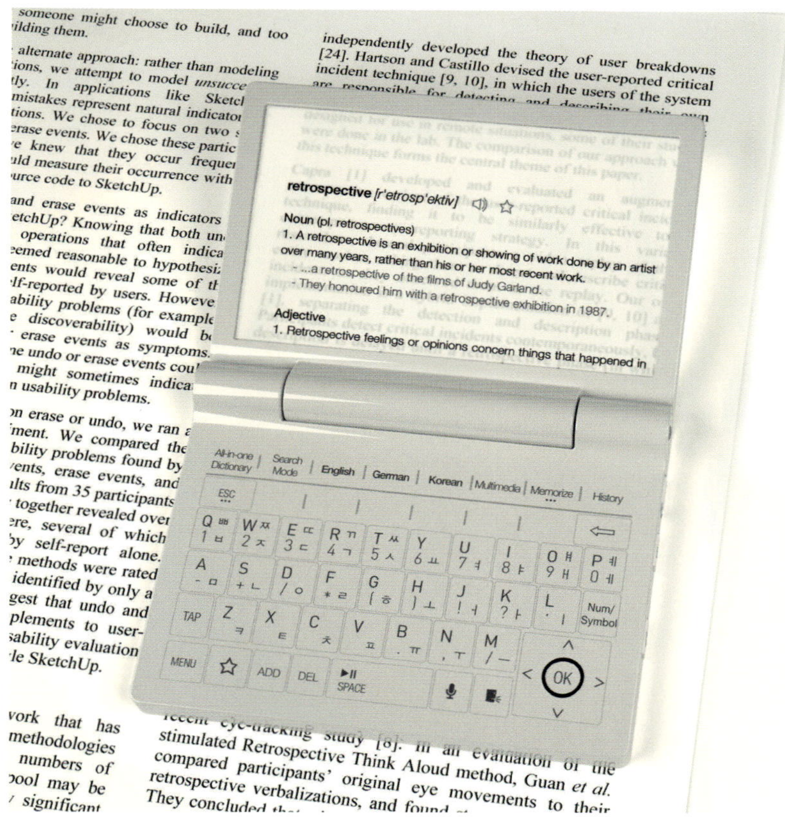

"The WYSIWYG dictionary is a good example of how transparent displays can be used to assist in learning 2nd languages intuitively. We want to take this opportunity to develop this concept further to be a full-fledged product."

AWARD WINNERS

The 80 iF concept design award winners 2012 in alphabetical order by the name of the entry. Their ideas reach from simple solutions to complex concepts.

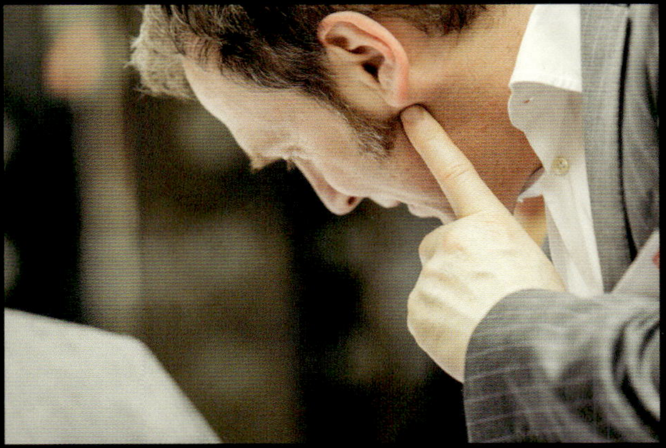

industrial design + product design

Design
Yu-Chia Nien, Chia-Yi Chou
Wei-Lin Kang, Ying-Chin Sung
Taipei, Taiwan

Da-Yeh University
Changhua, Taiwan

藥用OK繃
BAND AID

Medical Band-Aid is a band-aid with a disinfectant gel inside. When it is placed on the wound, the bubble of gel bursts and disinfects the area. The curved tip of the band-aid makes it much easier to remove.

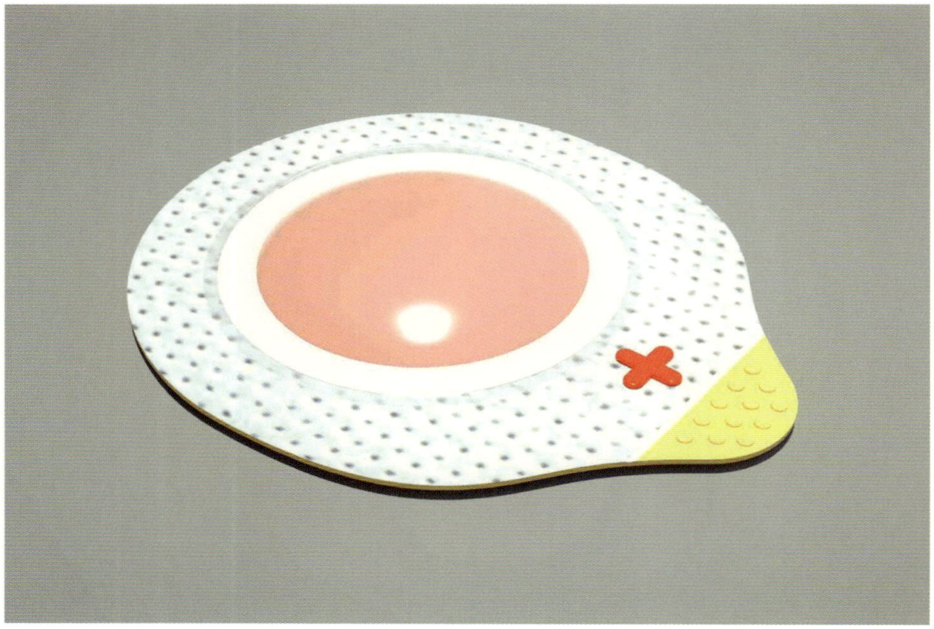

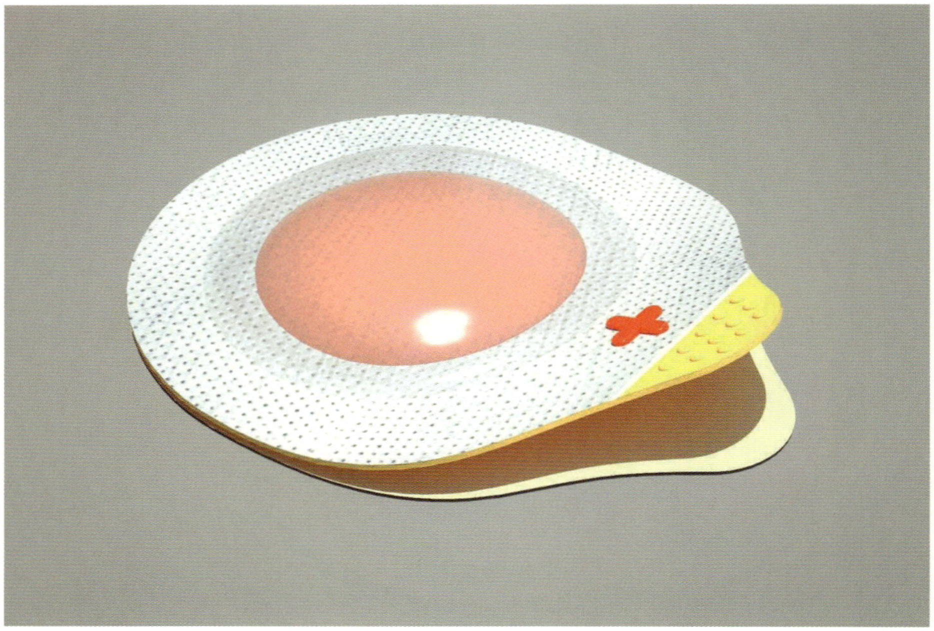

"Winning the iF concept design award 2012 is a positive response to our work as well as an honor for us. Thanks to our instructor, Chien-Ming Hsieh. Under his guidance, we were able to think of so many creative ideas and experienced good team spirit. Thanks also to our university and department for providing a platform of design for us to learn and to go beyond the normal limits."

industrial design + product design

Design

Chul Sung Ahn, Chang Dae Jeon,
Da Hye Lee, Hee Chul Kwak
Gyeongsan, South Korea

Daegu University
Gyeongsan, South Korea

0 POINT RULER
RULER

Conventional rulers normally have to be adjusted to the 0 point to then draw a line in order to be accurate. To eliminate this step, the 0 point ruler automatically goes back to the 0 point wherever the pencil touches it. The entire top side consists of a pressure-sensitive touch screen, which returns to the 0 point whenever it senses pressure. Using the 0 point as the starting point, it marks the numbers that go before and after.

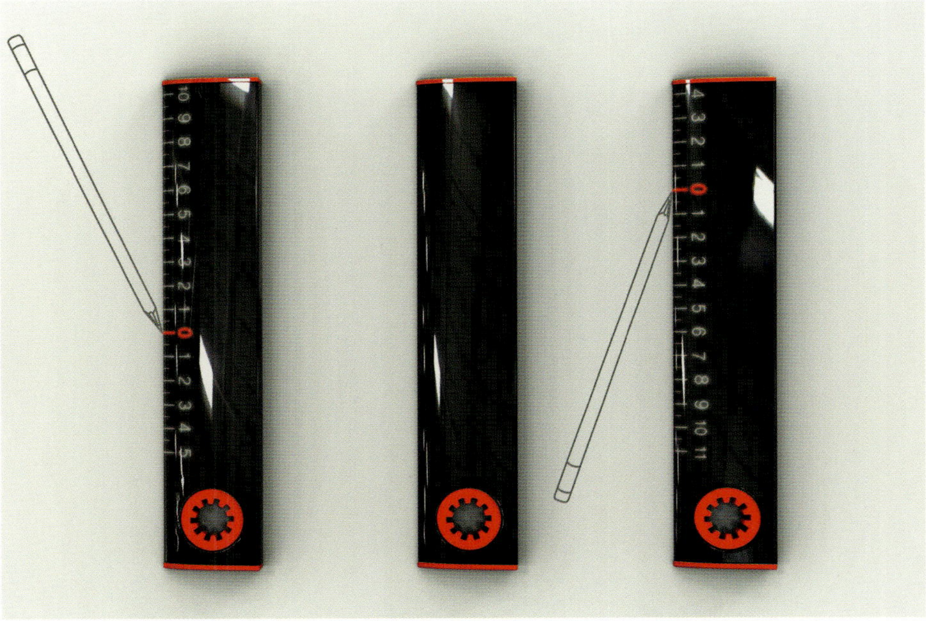

Leticia Castro Gaziri: *"This ruler would be an improvement for people who need to use it a lot. It's something simple but new and reflects the idea of technology combined with sustainability."*

"I feel so honored to win this award; it will make me work even harder. I want my designs to help people lead better lives. I want to thank my family and friends. And I want to thank the iF concept design award, for receiving it proved a very valuable experience."

industrial design + product design

Design

Xiaoguang Zhao, Diying Song, Jingwen Li
Shanghai, China

Hohai University
Shanghai, China

A MAGNIFIER LID
MEDICINE BOTTLE CAP

Many people in their senior years have to take any number of medicines to maintain their health or cure diseases. Most of them, however, share the common problem that their vision and their memories are declining. As a result, it is often difficult for them to distinguish between different medicines and to remember the frequency and the dosage. In order to deal with this problem, an ordinary medicine bottle lid has been turned into a magnifier lid, making it easier to read the instructions and other information on the bottle. It is also very convenient to carry and to use.

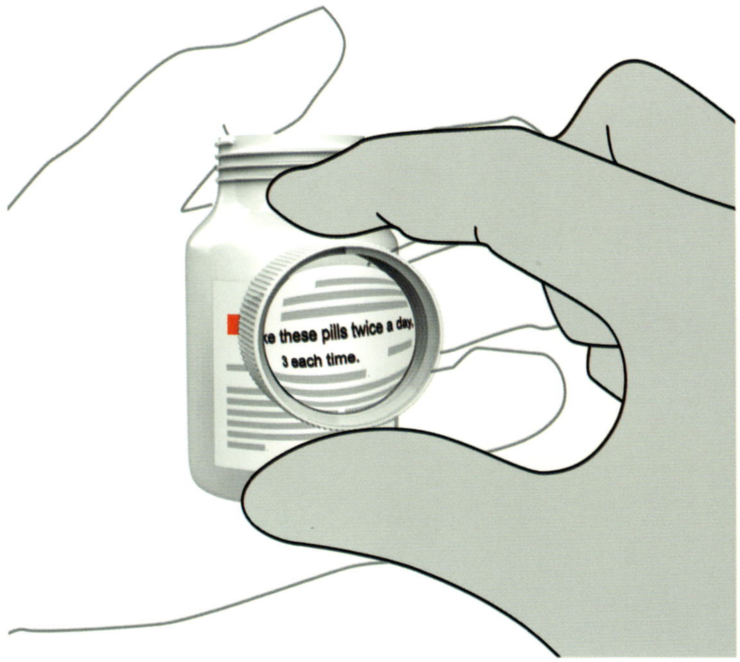

"Pay attention to the life quality of the weakness is the very duty of designers. Thanks to this competition that we learned the good design should come from the subtle aspect of our living behavior. What we've got is more than just a certification of iF, but a wholly new view of what designers mean to people, and what we should do to improve the life of others."

industrial design + product design

Design
Hao-Ming Hsu, Yi-Ting Chen,
Pek-Khai Tay, Cheng-Yu Tsai
Taipei, Taiwan

Shih Chien University
Taipei, Taiwan

A.C.R.
RESCUE EQUIPMENT

In an emergency, it is difficult to report and rescue quickly at the same time. During CPR rescue, the people who perform first aid are also often unsure of exactly how to do the cardiac massage: the strength, the depth and the frequency. Mouth resuscitation also carries a risk of infection. This new design makes professional first-aid equipment easy to use in all situations and environments and increases the success rate of emergency rescues.

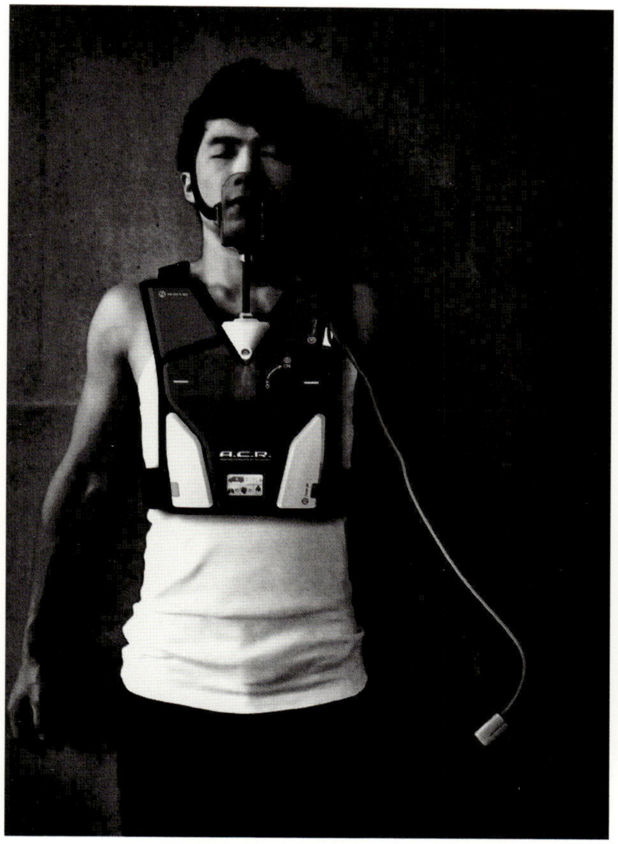

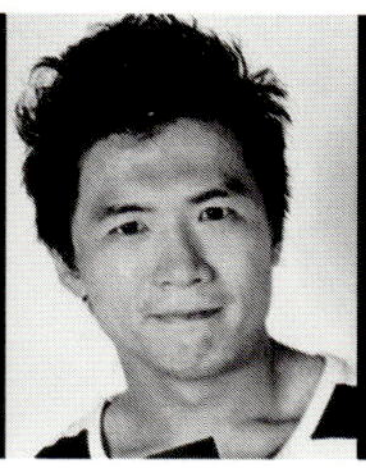

"We feel very honored to be granted this prestigious award. We are aware that it is one of the world's most renowned awards in the field of design, so it means a lot to us that iF appreciates our efforts and recognizes our achievement."

industrial design + product design

Design
Yakun Zhang, Yu Liu, Danwei Ye
Rochester, United States of America

Rochester Institute of Technology
Rochester, United States of America

ACOUSTIC-OPTICALLOCK
BICYCLE LOCK

Every bicycle needs a lock to keep it safe. However, a lock does not help to find your own bike on a crowded parking lot where there are usually hundreds of bicycles. Poor lighting sometimes makes finding your bike even harder. This acoustic-optical lock system uses a radio transmitter in the key and a radio receiver in the lock. When the button is pressed, the lock makes a sound and emits a light, making it easy to locate the bike. This is designed to make life a little easier for people who want to enjoy a low-carbon lifestyle.

Tiago Krusse: *"A helpful product that saves time and improves people's lives."*

"We are very glad that we won an iF concept design award concept design award and thanks for the whole iF community's appreciation of our works. Our group members have participated in the award on different teams for three years, and this is the first time for two of our group members to receive this honor. We are so happy not only because our hard work has been rewarded, but also because it represents a milestone in our careers."

industrial design + product design

Design
Myeung Kwan Ha, Hyunsu Kim
Incheon, South Korea

Hoseo Unviersity
Asan, South Korea

AIR BOX
TRANSPORT PACKAGING

The "Air Box" is a paper box with an air-filled vinyl lining designed to protect fragile objects when they are being carried or stored. The inside molds to the shape of the objects to prevent them from moving around inside the box and being damaged. This box is also very lightweight.

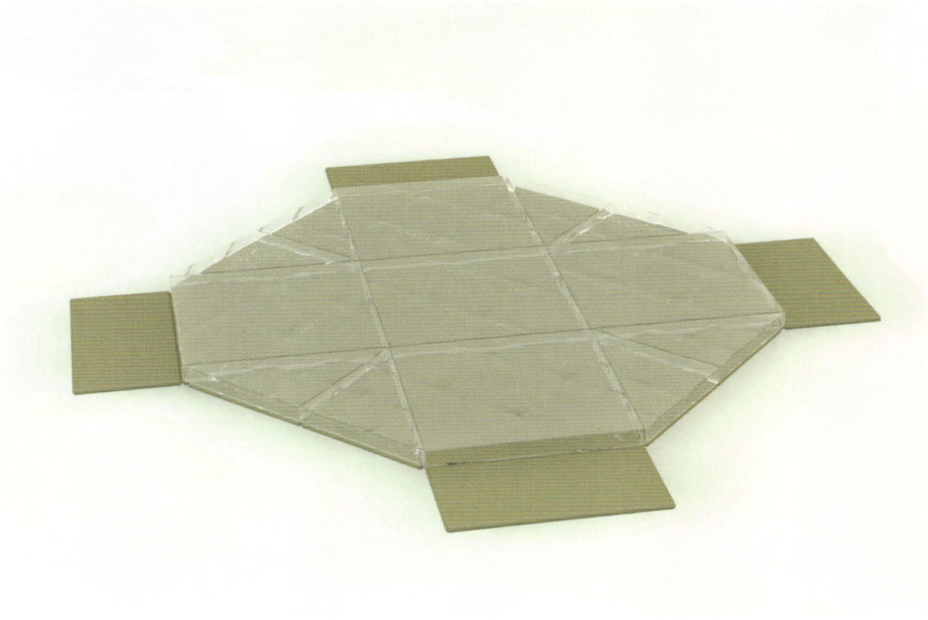

Elaine Ann: *"This concept is very feasible and solves a real practical problem elegantly, especially in today's e-commerce heavy world with billions of internet sellers."*

"Thank you. This is very unexpected, like starting at the top. A dream come true, I can hardly believe it. I hope my design will be able to make a difference. I look forward to my future as a designer."

industrial design + product design

Design
Hsin-Ya Huang, Kai-Lan Huang
Taipei, Taiwan

National Taiwan University of Science and Technology
Taipei, Taiwan

ALIEN EYE
LEARNING TOOL FOR KIDS

The "Alien Eye" is a toy that teaches children about such different topics as culture, people, animals and buildings around the world. It's made up of 2 parts: the observation device (the Alien) and a globe of the world. First, the children take the Alien and look at different places on the globe. The Alien reads the RFID code built into the globe and connects to the internet to search for the corresponding images, which are then displayed on its screen. Voice output provides key information about the image. This is an excellent opportunity for children to learn more about the world's geography and culture.

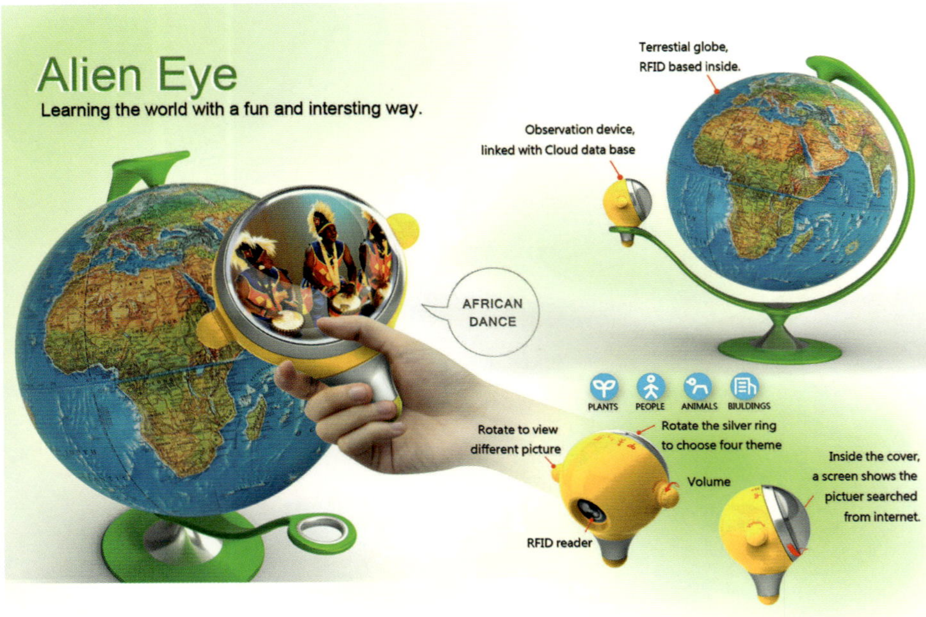

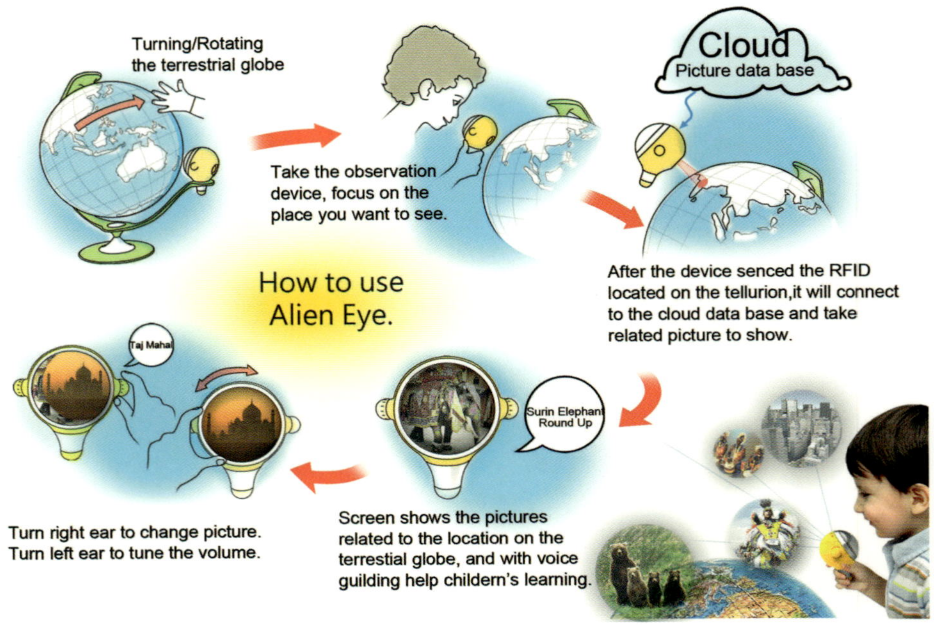

Leticia Castro Gaziri: *"Not just the idea is very interesting but it's an education concept that would allow children to understend other cultures and link them with their own countries, providing giving a good background in geography and cultural understanding. Although still only in the concept phase, the project is well done."*

"Thank you, iF, for giving us this opportunity. We still have a long way to go."

industrial design + product design

Design

Shijian Luo, Guotao Hou, Jiaqi Pan, Yi Hu, Wenjie Li
Hangzhou, China

Zhejiang University
Hangzhou, China

ALWAYS TOGETHER
WATCHES

"Always together" are twin watches designed to keep you close to the people you love. They use GPS technology and can easily help you to find your partner's location and distance from you. The big cam of one watch always points towards the other one. And the little cam displays the distance between the two watches. The distance is divided into 4 parts, in each of which the little cam moves evenly. The closer you get, the slower the small cam moves.

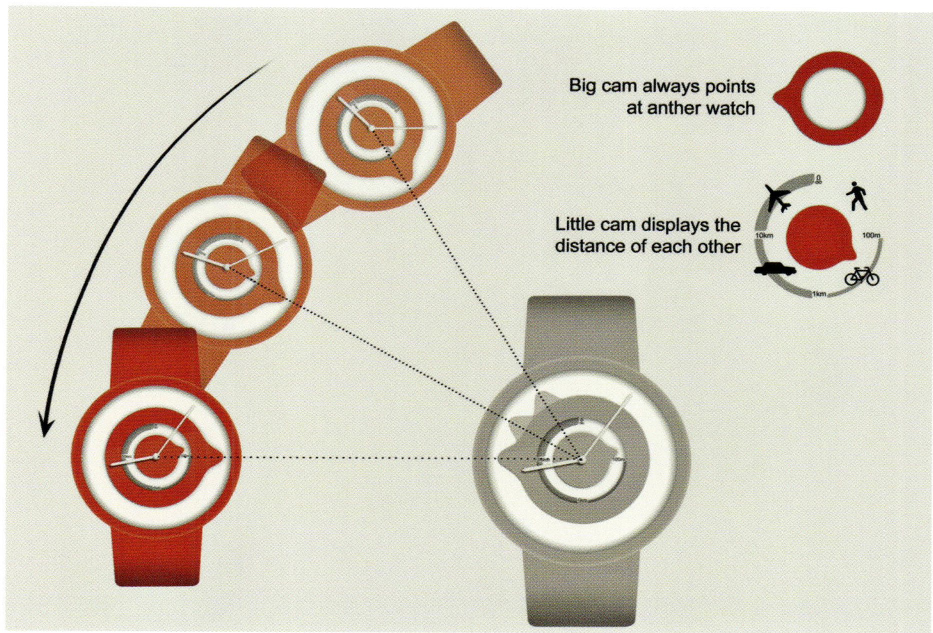

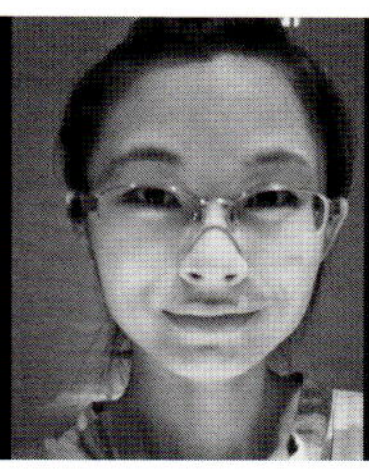

"It is a great honor to receive this special prize, which is also proof of our efforts, and we believe through the well-developed and efficient work processes, the FIDT can bring good design to clients with research, creativity, teamwork, and international resources."

industrial design + product design

Design
Chul Hee Lim, Hyun Woo Yoon
Ansan, South Korea

Hanyang University
Ansan, South Korea

AMASSTATION
GAS TRANSFER DEVICE

The "AMASStation" is used to collect leftover butane gas from cans that can no longer be used because of lack of pressure and injects this gas into another container so that none is wasted.

Lidan Liu: *"With respect to collecting neglected remaining gas, this concept reflects a measure of social responsibility. It is a useful idea. The design is self-explanatory regarding function and exhibits clear product identity. It has also achieved a certain aesthetic appeal."*

"We were able to successfully finish this project in a short period of time because we helped each other and shared a common goal. This prize is very special to both of us, and we will continue to learn more and develop better product designs for the future of our world."

industrial design + product design

Design
Kun Li, Fuwei Sun, Qian Li, Shizhong Fan
Changsha, China

Hunan University
Changhsha, China

ANGLE EYES
CAP OF MEDICAL BOTTLE

"Angle eyes" is a bottle designed specifically for the elderly or the visually impaired. Simply remove the bottle cap and then use it to magnify the information on the bottle. This can effectively prevent serious mistakes in taking medication.

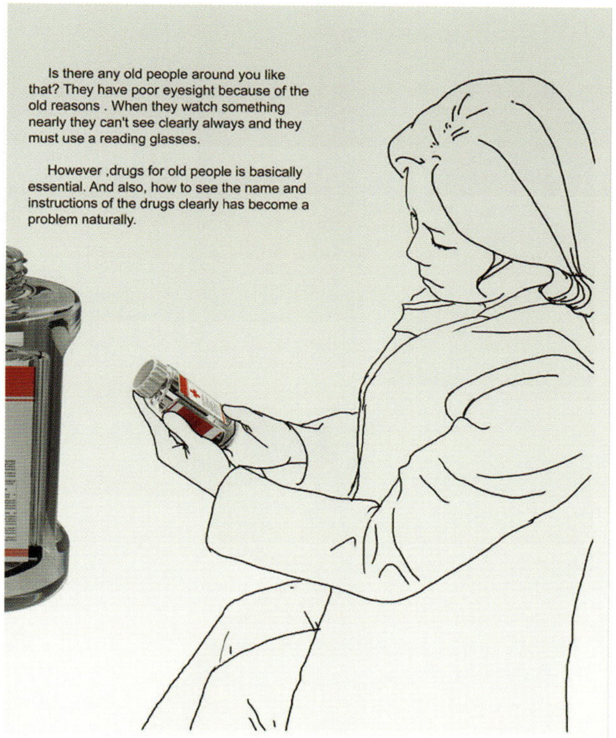

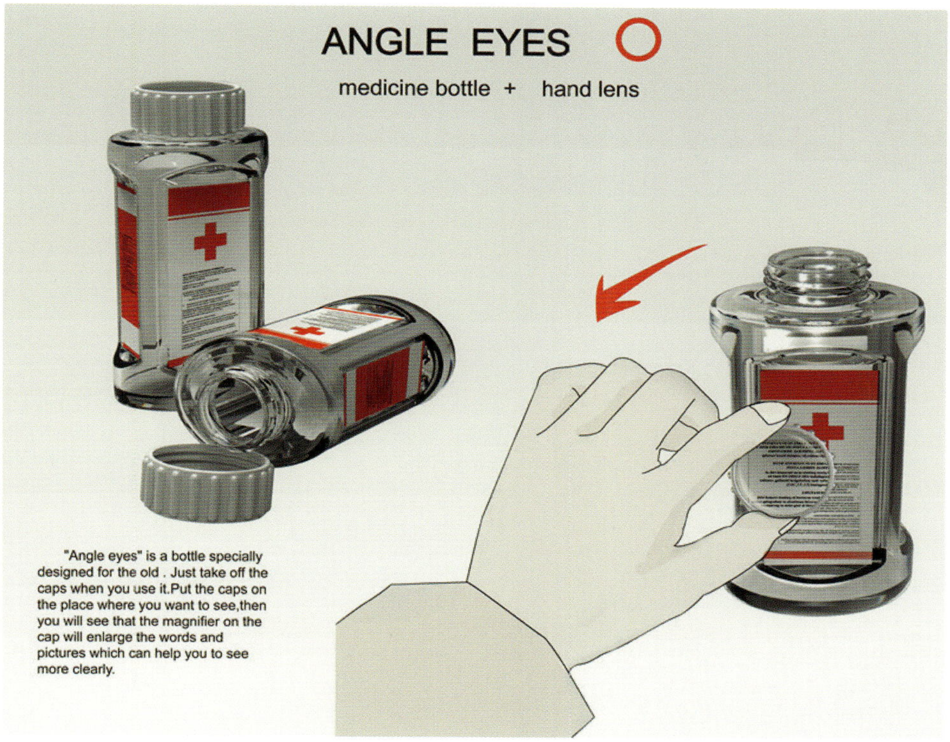

Rolf Schweizer:
"Idea very creative. Presentation very good."

"We feel honored to accept the iF concept design award, a great achievement for us as designers. We are grateful for this opportunity to showcase our talents and for your full endorsement. This has been great life experience. We will try our best to make a little difference to our daily lives, now and in the future."

industrial design + product design

Design
Fan Wu
Guangzhou, China

Hubei University of Technology
Wuhan, China

ANYWAY SPRAY
PRODUCT DESIGN

It is often difficult to get the last few drops out of a spray bottle. This bottle uses the force of gravity to ensure that the suction mouth is always kept below the level of the liquid contents.

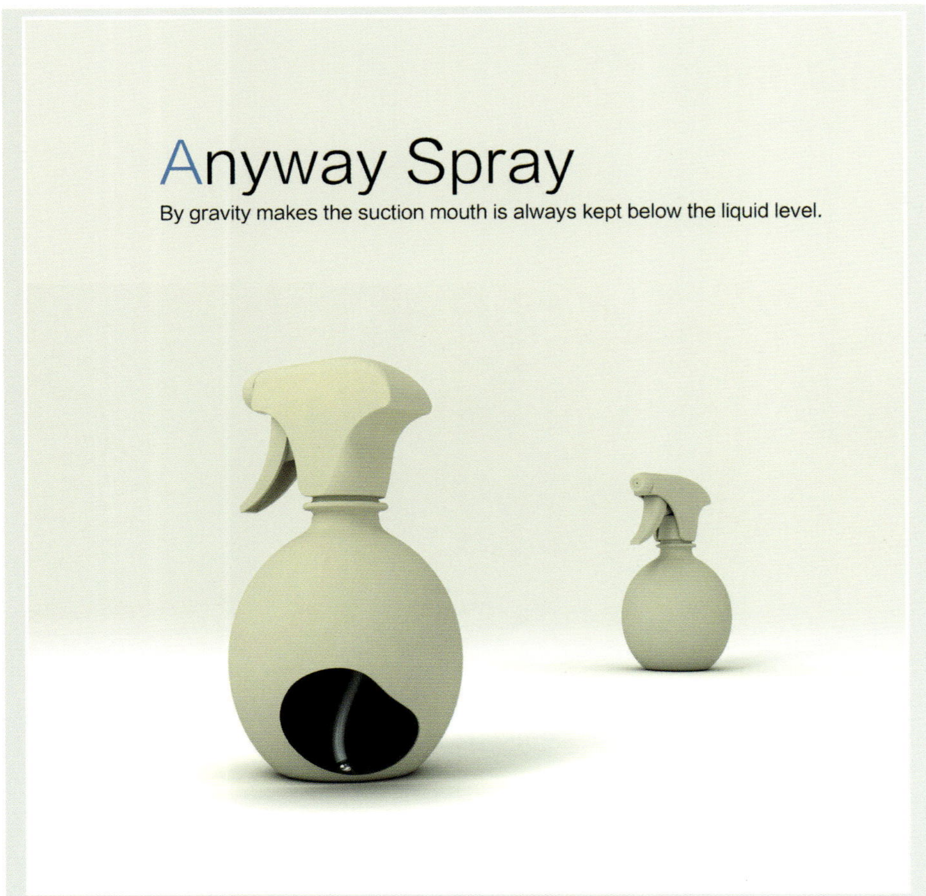

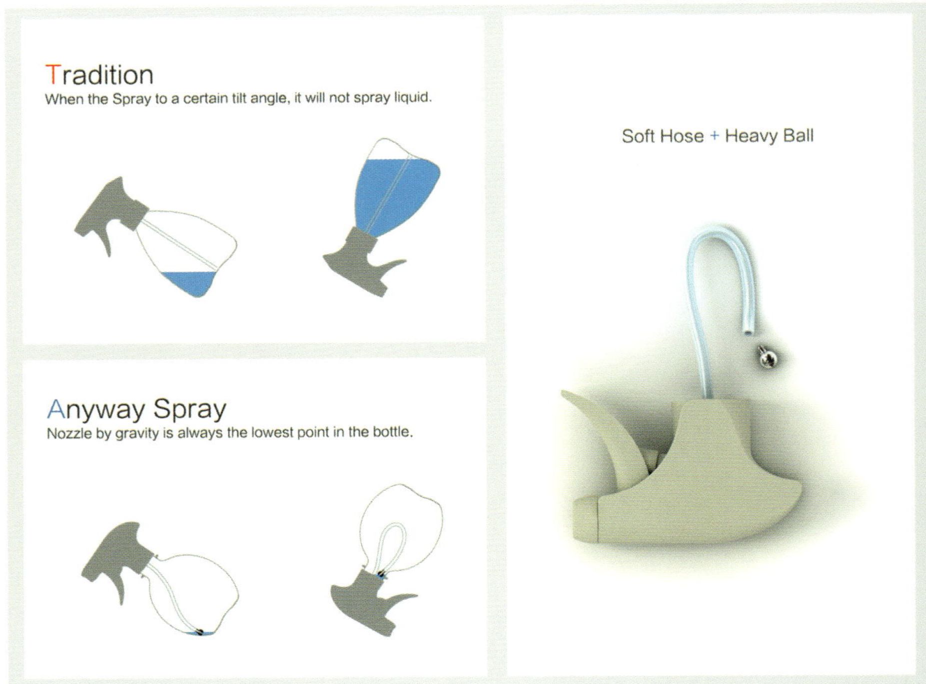

Adrienne Viljoen: *"The entry impresses as it addresses a common problem successfully by making use in an innovative way of simple gravity principles. It is very easy to implement and will be compatible with many products. The product is self explanatory and the presentation clear and attractive. No material has been specified."*

"I am honored to get this award. iF taught me how to find ways in which to improve life in general. It has given me design insight and a new understanding of design, I will continue to work hard."

industrial design + product design

Design
Maojin Ye, Haohao Wu, Ruili Sun, Nan Wu
Nanchang, China

Jiangxi Universit of Finance and Economics
Nanchang, China

ARROW CARRIGE NUMBER GUIDANCE SYSTEM

When boarding a train, it is often difficult for passengers to find the train compartment and seat they have reserved. The "arrow carriage number" is the solution to this problem. The middle number on the board indicates the compartment, while the two side numbers show the direction either forwards or back, so that passengers know which way they have to go.

"We are honored to receive the iF concept design award. We will continue to work hard. First of all, thanks to my parents for their support. Secondly, thanks to the iF platform; it is a great opportunity for all young designers to display their talent. Thirdly, thanks to everyone on our design team; the secret of success is team work. Finally, I believe everyone is able to take a small step to make this world a better place."

industrial design + product design

Design
Ya-Ting Yu, Hung-Ming Chen (Tamkang University)
Taipei, Taiwan

National Taiwan University
Taipei, Taiwan

BAMBOO KNOT CHOPSTIC
CHOPSTICKS

These chopsticks are made of bamboo with the original knots in that material. The uneven surface on the chopsticks means that the tips that touch the food and the user's mouth do not rest on the table when they are put down, keeping the immediate area more hygienic and clean. Compared with plastic or medal chopsticks, bamboo is more flexible and has an additional anti-bacterial quality.

Simone de Waart: *"It is a nice, simple, but effective and aesthetic idea and still remains close to the traditional chopsticks."*

"We appreciate very much the efforts of iF and the jury to remind young designers to design works with a more practical and modest attitude. This is a process filled with fun and challenges and will undoubtedly cheer us on to continue to think of and develop feasible, useful and sustainable designs in the future."

industrial design + product design

Design

Yi Liu, Chao Cui, Lei Sun, Jialan Zhou
Taiyuan City, China

Xi'an Technological University
Taiyuan City, China

BLIND TAPE
BRAILLE TAPE

This tape is a new type of tape that has been specially designed for people with poor eyesight and for the blind. The protruding dots on the tape can be pressed to help solve such everyday problems as classifying objects, leaving messages, providing information, marking objects, etc.

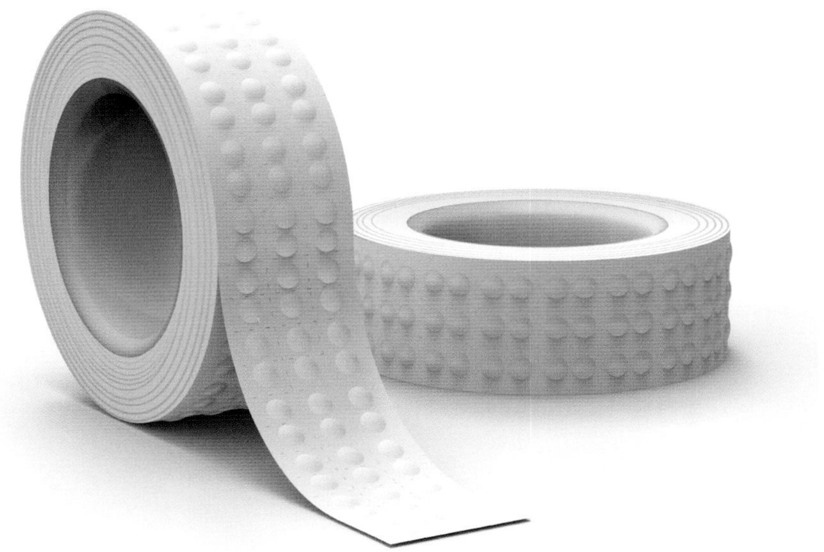

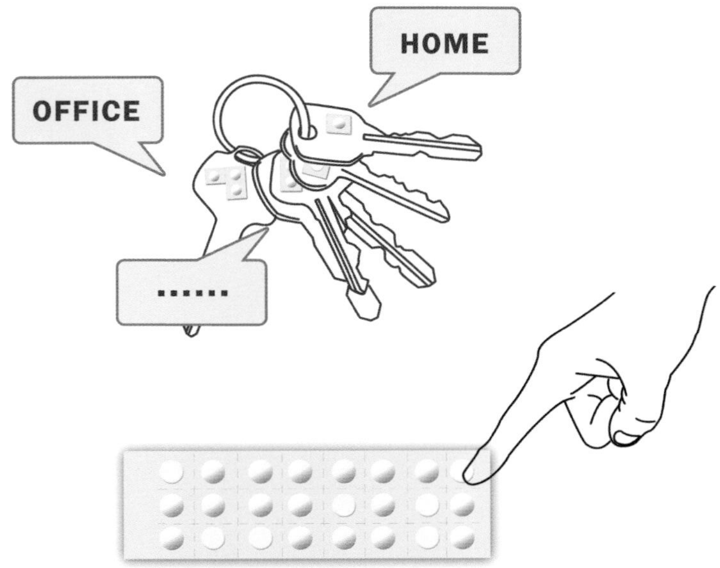

Jan Stavik: *"An excellent user-orientated design for the entire concept."*

"Thanks to iF for providing this opportunity to students. As young designers, we enjoy the process and it encourages us to try out more types of design. This is a wonderful way in which to improve our understanding of design."

industrial design + product design

Design

Shu Wei-Chung, Wan-Chi Chang
Taipei, Taiwan

National Taiwan University of Science and Technology
Taipei, Taiwan

BLINDJOG
SPORT PRODUCT

"Blindjog" makes it possible for the visually impaired to jog by themselves. It is made up of three main parts – earphones, two anklets and sensor plates, which make sure that visually impaired stay on the track and are heading in the right direction. The system uses RFID technology and Bluetooth technology.

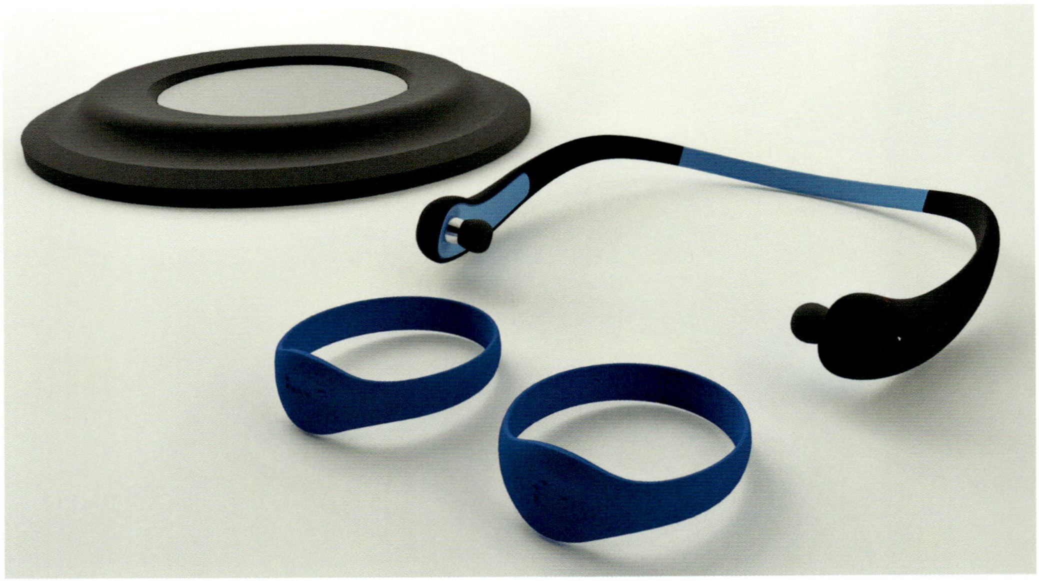

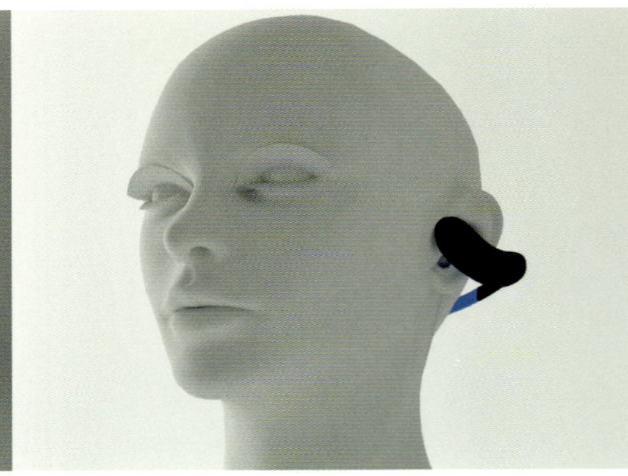

„Being encouraged by winning the iF concept design award leads us to further research and innovate creations for the visually impaired. Thanks to those who supported our research and hard working during the project, especially my parents, professors and my team member! It's a great surprise to win as it's our first time to have entered this contest. We will keep up being passionate about design."

industrial design + product design

Design

Jieun You, Hyewon Kim, Younsung Lee (Kunkuk University), Gyujung Lee (Sungshin Women's University)
Seoul, South Korea

Kookmin University
Seoul, South Korea

BLISTER CONTAINER
CONVINIENCE FOOD PACKAGE

Ramen noodle dishes often include separate spices and seasonings, which can be added as desired. These small packages can be difficult and inconvenient to tear open and sometimes the top of the container is damaged as well when it is opened to remove the seasoning. This new blister Ramen container solves the problem. You simply press down on the blister and the seasoning is dispensed into the container.

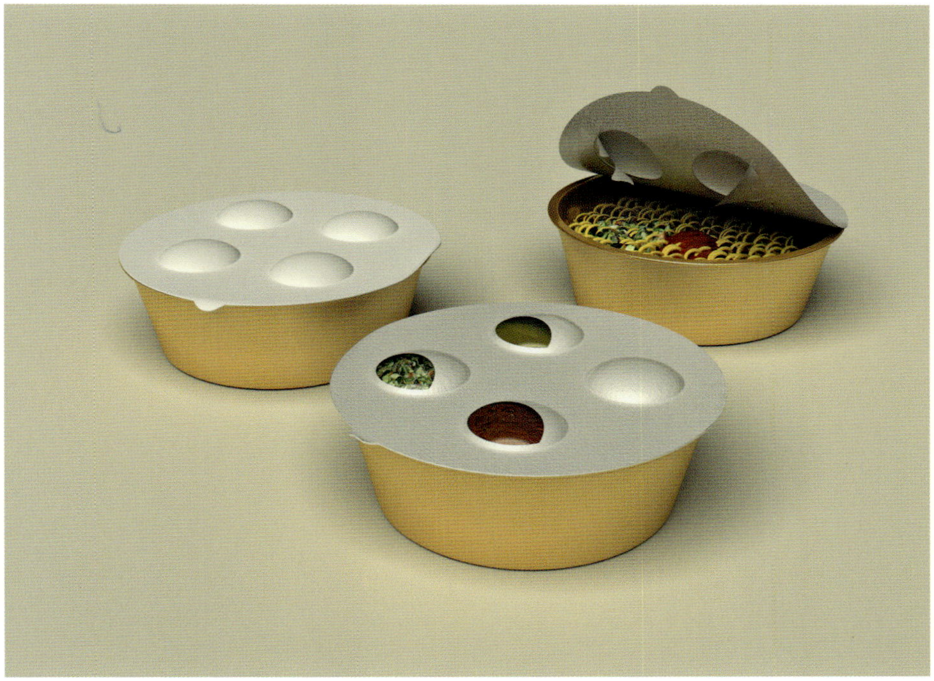

Li Bei: *"This is a creative design. The designer improved the original ramen container after observing life in order to meet the target customer's requirements. The new design is convenient to use, sanitary and personalized. It would be easy to manufacture. Using this kind of design can reduce packaging materials and garbage, corresponds to the requirements of saving resources and environmental protection."*

"We can't believe till now that we became winners! We are happy to be winners."

Design
Kuang-Ching Tai
Tainan, Taiwan

National Cheng Kung University
Tainan, Taiwan

HEAT CONDUCTION BOWL
BOWL

When pouring liquids, the blind or the visually impaired usually put their fingers into the container to know when the liquid reaches the desired level. For very hot liquids, such as tea, however, this is not an option. This bowl has been designed using heat-conducting bars so that it is possible to feel the level of hot liquid in the bowl by touching the bars.

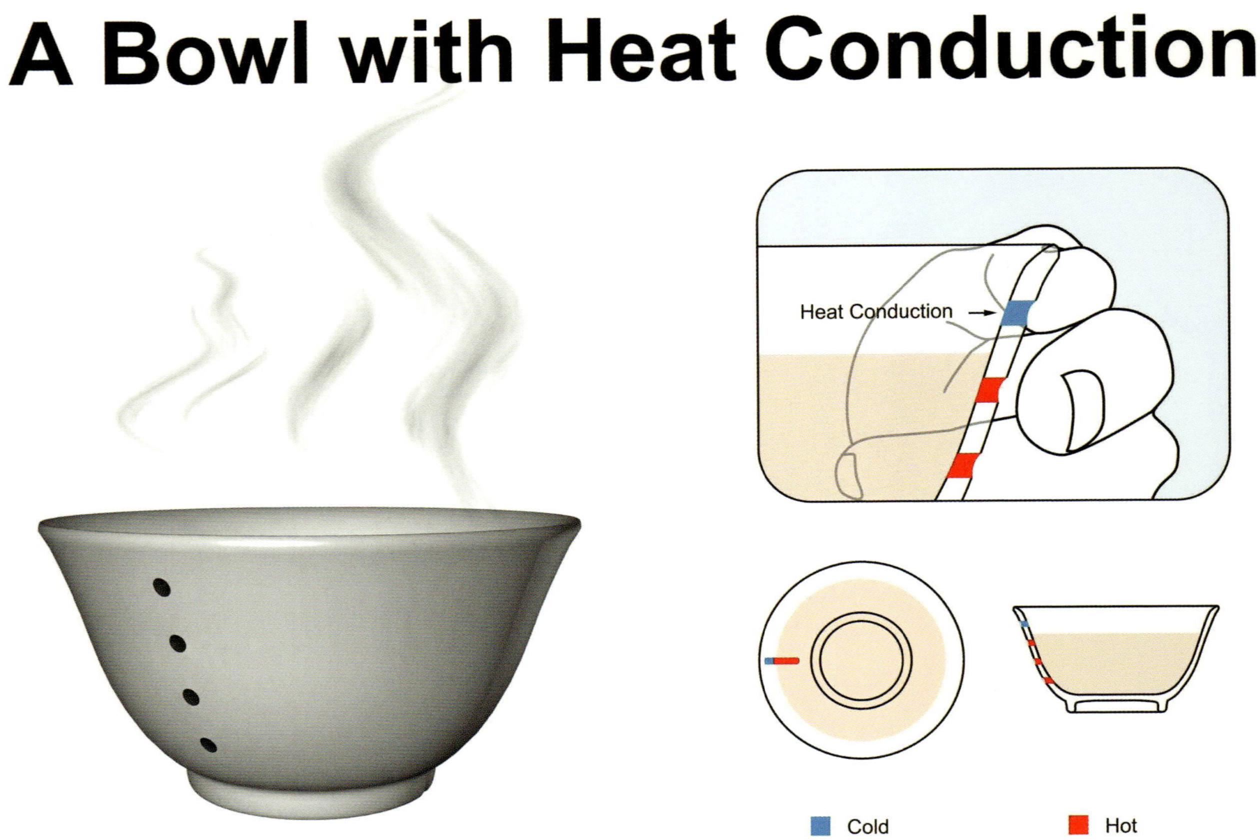

Szonja Szesztai: *"Simple and elegant design, easy to use. Further uses can be still developed."*

"I am amazed to be among the award winners! I prayed to God for his favor. To me, this prize means that I have greater responsibility to design products to assist people who need to be assisted."

industrial design + product design

Design
Yen-Chih Chen
New Taipei City, Taiwan

Chang Gung University
Tao-Yuan, Taiwan

BRAILLE LABEL
CLOTHES LABEL

According to the WHO, there are 285 million people worldwide who are visually impaired. Incapable of seeing colors or sizes, they are virtually unable to dress themselves the way they want to. This concept suggests printing Braille information on the clothing label to give the blind greater independence in dressing themselves.

"I hope the braille clothing label will help the blind to feel more independent. I am glad that this concept was appreciated by the judges and feel encouraged that my research has been able to make a contribution to the design industry."

industrial design + product design

Design
Soonkyu Jang, Jieun Seo
Gwacheon, South Korea

Dankook University
Yongin, South Korea

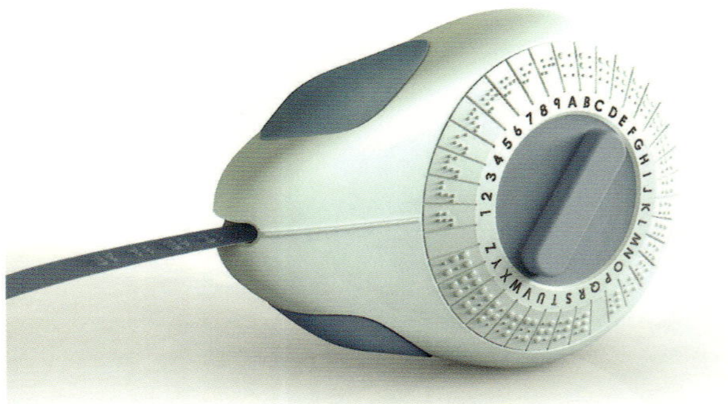

BRAILLE LABEL
BRAILLE TAPE

Braille is the written language of the blind and provides them with information on the world around them. The rest of the world, however, has little experience with this language. That's why we invented the "Braille Label", a tool that is designed to make communication between these two groups easier. Seeing people can produce messages in Braille using the regular alphabet, while the blind can produce sticker using Braille.

"We are happy to receive the iF concept design award this year. We love designing together at university. Even though one of us is a visual designer and the other is a product designer, but the major is not so important, since what really counts is working together in the field of design. Thank you for giving us the iF concept design award!"

industrial design + product design

Design
Jiheon Kim
Seoul, South Korea

Seoul National University
Seoul, South Korea

CANDLE LIGHT
LED LIGHTING

This product has been inspired by candles; it provides the same emotional feeling and the glow, but there is no open fire involved. The light goes on when you touch the top of the candle-like device and you can adjust the intensity of the light by simply touching it again. A built-in timer automatically switches the light off after a pre-programmed period.

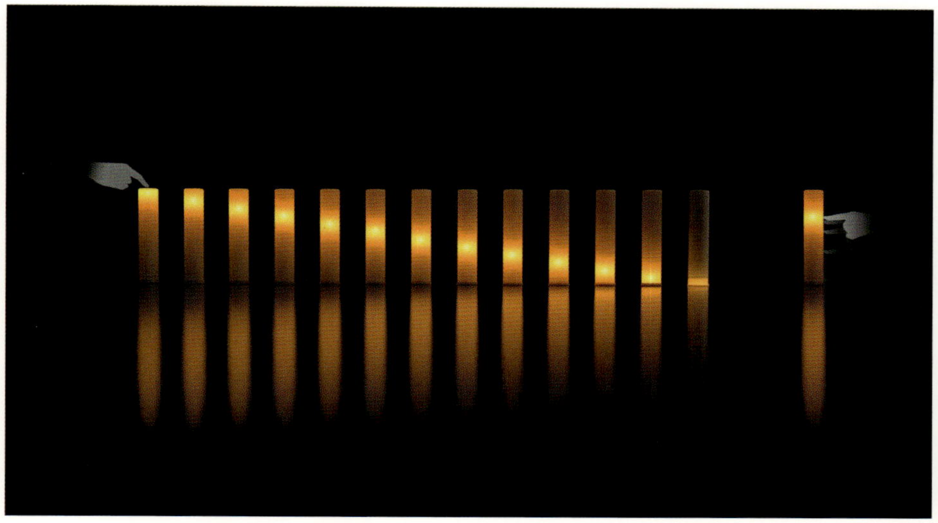

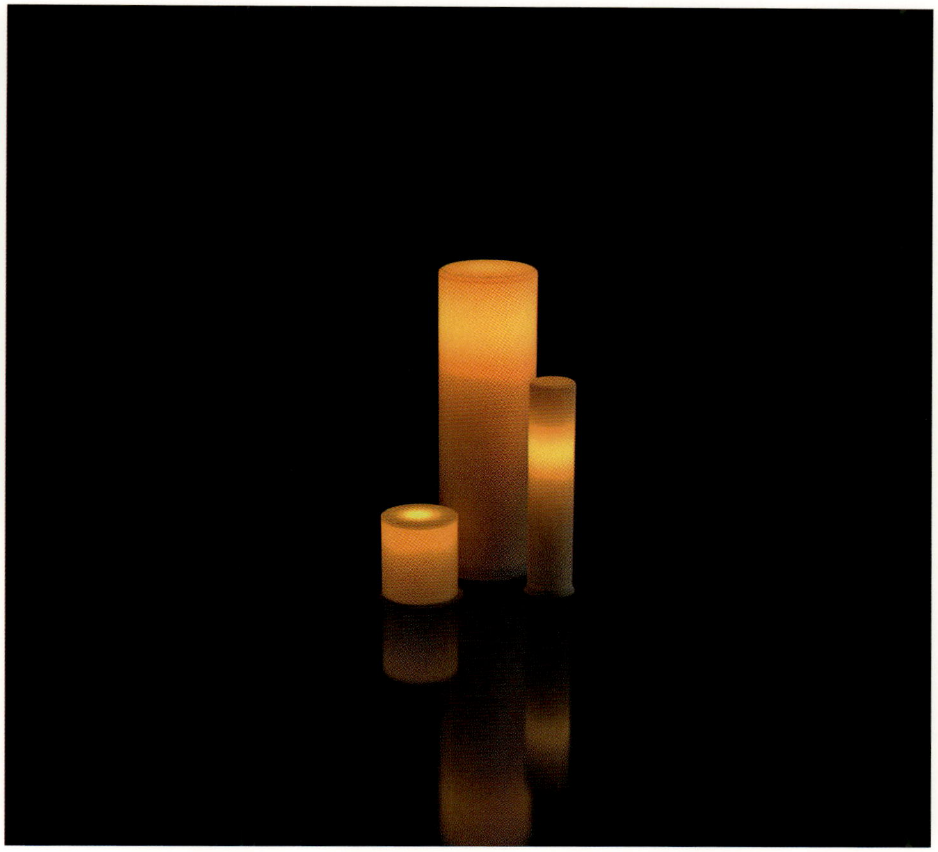

Leticia Castro Gaziri: *"The idea of representing a candle and the passage of time has brought an interesting feeling to the technology product. There is one aspect of wellness in the solution that leads to the house, the pleasant moments. Technology and the formal solution are well-worked results in a distinctive design."*

"First, I'm very thankful to receive this award. This has been immensely rewarding. And I shall see you some time again as a product designer. Not just a student. Thanks again, and I wish you all the best."

industrial design + product design

Design
Chen Jinxin, Wang Xiaoxue, Yi Ying
Shanghai, China

East China Normal University
Shanghai, China

COLORFUL MARKER
MARKER

Sometimes you have to go through a lot of different pens to find exactly the shade of color you want. This is time-consuming and inconvenient. "Colorful Marker" uses the principle of adding water to adjust the color depth. Each pen is dedicated to one main color and you simply turn the pen to get the ratio of water and color that you want. This greatly reduces the number of pens you have to carry with you and puts a virtually unlimited number of shades of color at your fingertips.

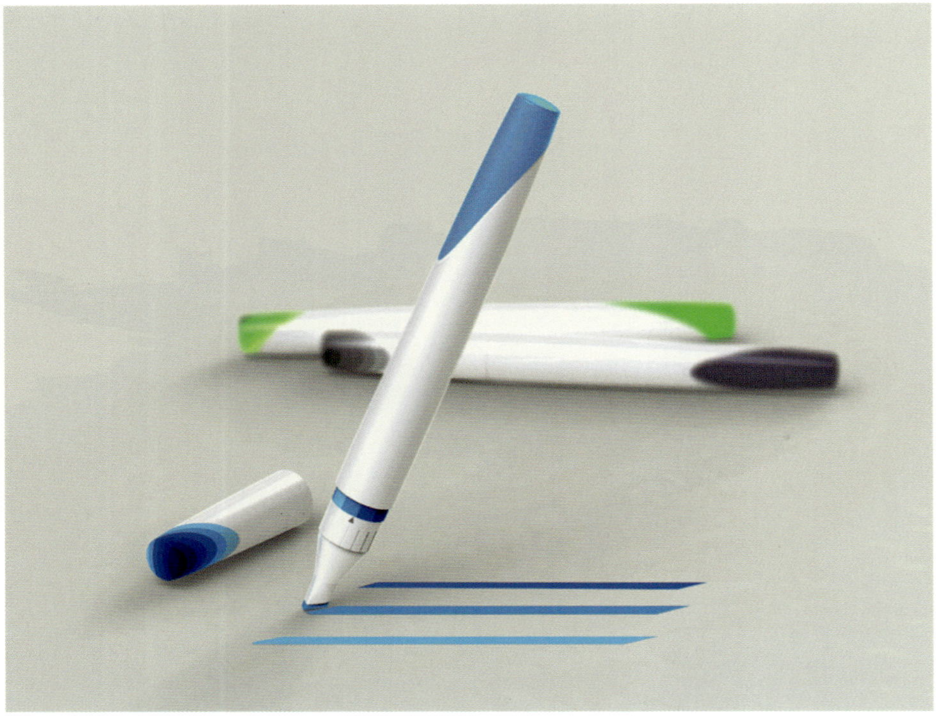

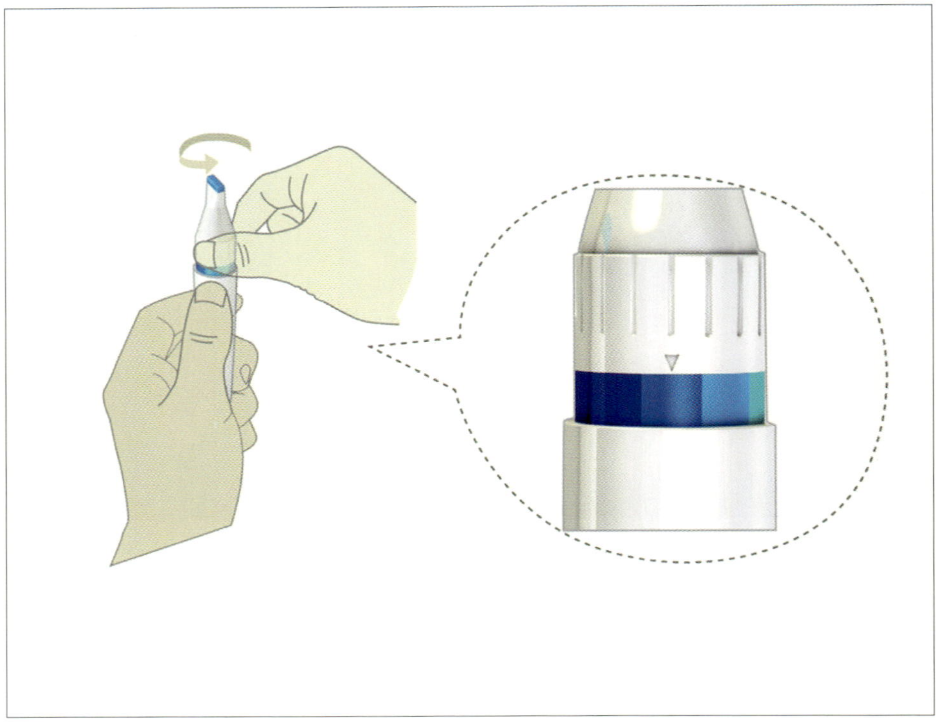

"We greatly appreciate iF for their recognition and for giving us this award. We owe our success not only to our own efforts but also to the support and assistance from our college. Facing both opportunities and challenges, we believe that we can march forward courageously and achieve more in the field of design."

Design
Cleber Niels, Rodrigo de Freitas Wolf
Curitiba-PR, Brazil

Pontifícia Universidade Católica do Paraná – PUCPR
Curitiba-PR, Brazil

CONDENSADOR
VETERINARY EQUIPMENT

This project is a special veterinary tool designed to diagnose a horse's respiratory system. The product collects information on the respiratory system by condensing the gases exhaled in the horse's breath to create liquid samples that can then be sent in for analysis. When the animal breathes, the inhaled gases go through the whole organism, passing through the lungs and the bloodstream. When exhaled, these gases provide information on the entire body. This information is collected using a non-invasive, fast, clean and cheap method, quite unlike the current, more complicated methods of diagnosis.

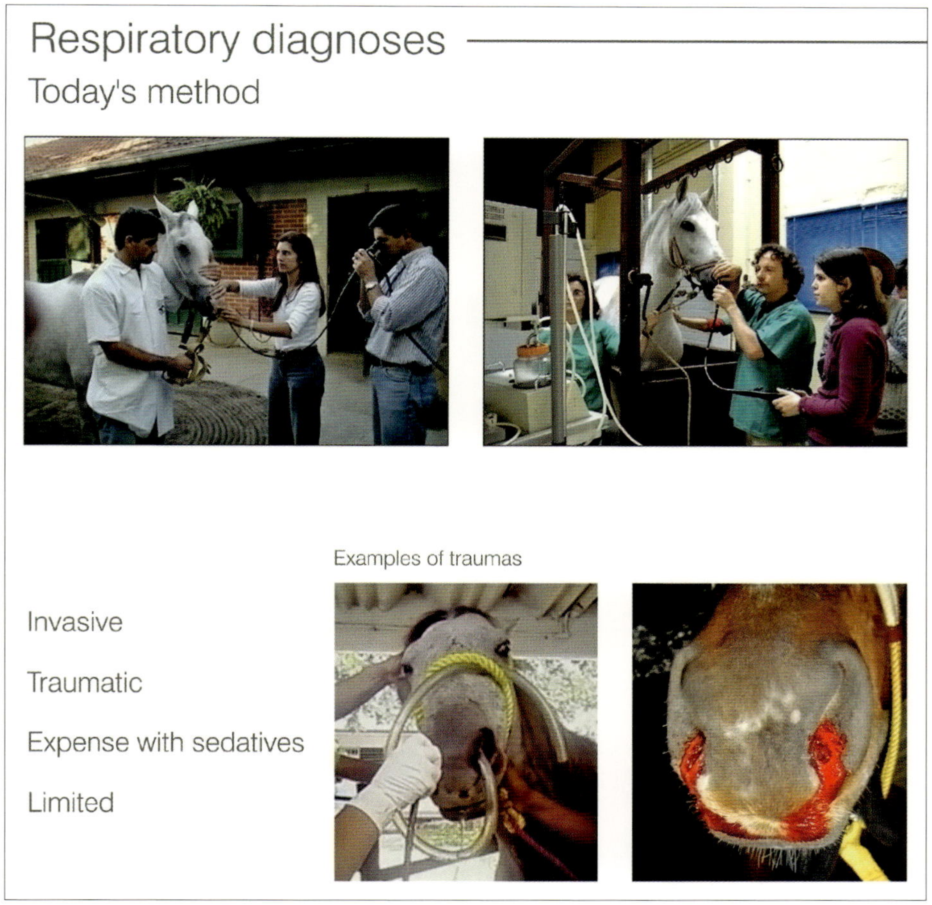

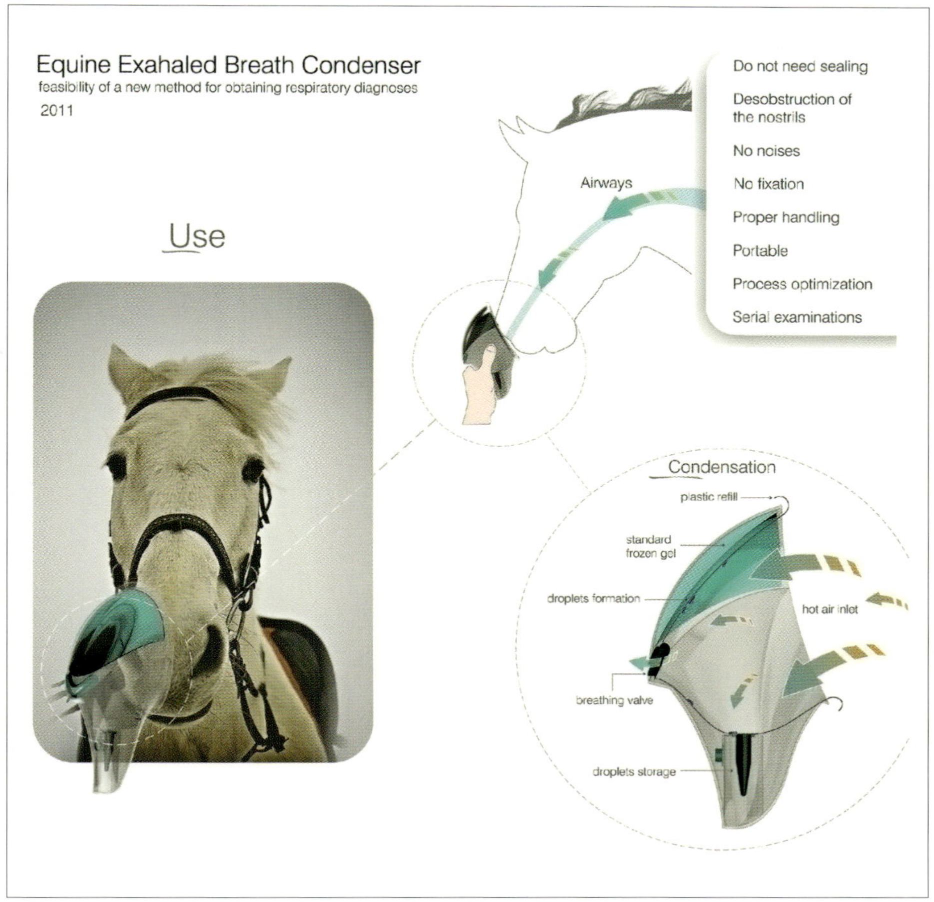

Prof. Martin Topel: *"Interesting concept for treating horses that is simpler than any one I have ever seen before. Technically, the concept appears mature and well-thought-out."*

"We think that winning an iF concept design award is the highest point in a design student's life. It's a very awesome beginning that can open a lot of doors. It brings opportunities that are very hard to get if you aren't an award winner. It's a dream come true. We want to say thank you to our family and all the people who helped us (they know who they are)."

Design
Yi-Chia Lee, Kuo-Chieh Wang, Shyh-Chung Chung
Taipei, Taiwan

National Taiwan University of Science and Technology
Taipei, Taiwan

CONNECTION CAP
PLASTIC BOTTLE CAP

Disposable plastic bottles take up a lot of space and are not biodegradable. This type of trash can be found on beaches in particular. While the bottles are often picked up, the small plastic caps get left behind. This can be dangerous for birds and fish. "Connection Cap" uses a screw-shape to keep the cap attached to the bottle. The bottle and cap can be held in one hand and then get tossed into the recycling bin together.

"Our thanks go to the jury first of all. During our preparation for this contest, we learned how to make use of each other's expertise and share the workload efficiently. This award means a lot. It confirms Taiwan's design strength and we greatly appreciate the international recognition."

industrial design + product design

Design
Kai Ming Yang
New Taipei City, Taiwan

Tatung University
Taipei, Taiwan

Adrienne Viljoen: *"This entry impresses as it addresses a universal problem in an inexpensive and aesthetically pleasing way. It is solved by using the universally understood symbols of green and red. There is no safety risk and very easy to implement."*

CONTACTLESS DEVICE

When leaving a public toilet, people often don't want to touch the door handle because you never know whether it's clean or not. This handy little device makes it possible to open the door without having to touch the handle.

"Dreams come true! It's a great honor to be one of the award winners. This title not only encourages me, it also gives me the confidence I need to keep working in this field. In short, thanks for your affirmation!"

Design

Cornelius Comanns
London, United Kingdom

Hochschule für angewandte Wissenschaften – Fachhochschule München
München, Germany

CONTAINER
KITCHEN TOOL

When adding salt or spices to your foods, it is very practical to use your fingertips because this gives you perfect control over how much is added and where. This newly designed container incorporates this "fingertip" feeling. It fits snugly into your hand while you gradually open or close your fingers to release the contents as desired. The silicone skin where your fingertips are placed is very thin to provide haptic feedback on how much of the contents have been released. This container allows the user to spread salt, spices, etc. in a very intuitive way, is more hygienic and can be used with only one hand.

Julian Lindley: *"The first impression is important and each time I opened these files I felt the visual object was "right". A good understanding of materials allowing a fresh approach in design. Project has been clearly and simply communicated."*

"If you invest much time and energy into a project then of course you are hoping to be rewarded. It's of course very nice you receive this recognition in form of winning an iF concept design award. Naturally, the project itself should be convincing, which means it needs to be noticed in the masses of new products. An iF concept design award can really help with that."

industrial design + product design

Design

S. Dogan Sekercioglu, Gosha Galitsky, Carol Tang
Umeå, Sweden

Umeå Institute of Design
Umeå, Sweden

DDSS
DIVER DISTRESS SIGNAL SYSTEM

A problem can quickly develop into an emergency in an underwater environment and once it does, every second until the diver is rescued is crucial to that diver's survival. The "DDSS" provides for fast and easy distress communication, making all members of the diving group instantly aware of the emergency that a single diver is experiencing. Specialized alarms and light signals are used to communicate various levels of distress. Once activated, the "DDSS" functions as a beacon that guides the rescue divers to the victim, saving precious time and significantly increasing the victim's chances of survival.

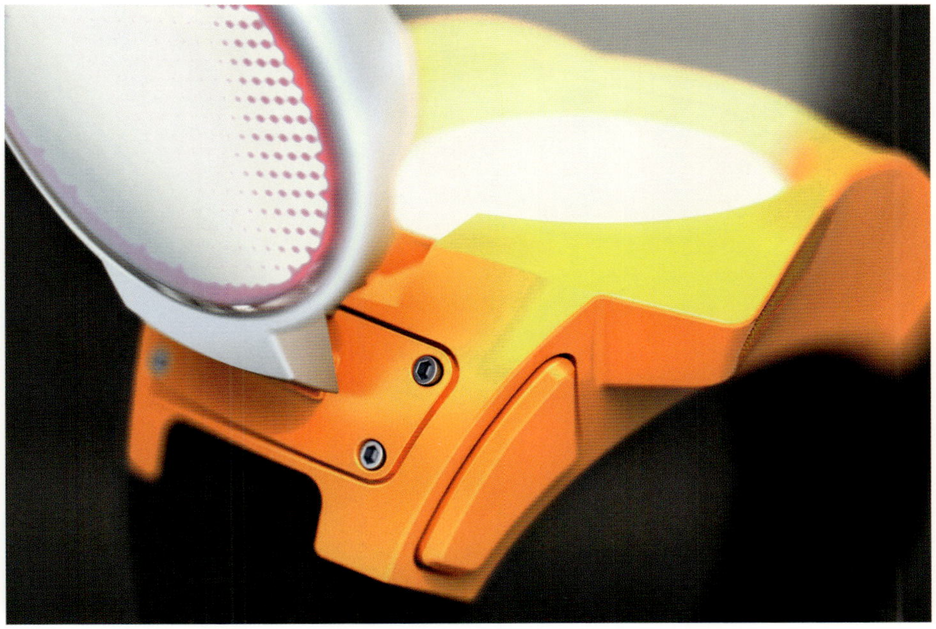

"Designing the Diver Distress Signal System was a multi-disciplinary team effort with the goal of creating a well-thought-out product that would address an important real-world problem. We are very excited and grateful for the iF jury's recognition since it means that we have achieved our objective."

Design

Jan Weber
Trier, Germany

Fachhochschule Trier
Trier, Germany

DO FURNITURE / TABLE
PARAMETRIC TABLE

"do furniture" is a parametric furniture system based on intersections or "nodal points" made from standard steel tubes laser-cut to shape. The form of the furniture is not fixed but rather defined by a set of rules. Specific parameters can be adjusted to produce the final geometry of any one piece.
Entirely translated forms can then be created according to the customer's individual preference. The exact profiles of these nodal points are calculated for precision production by a 3D laser cutter. Similar to a 3D-puzzle, these complex geometries are then assembled and welded together, which allows infinite variety in terms of form and design.

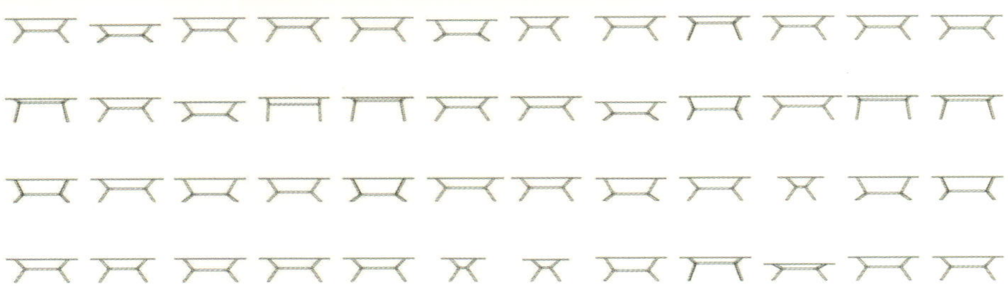

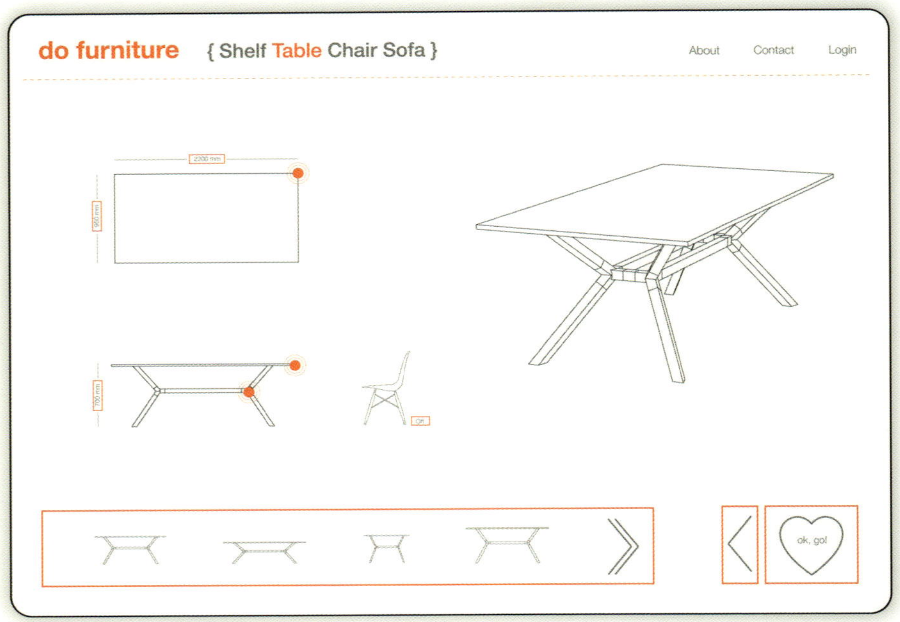

James Irvine: *"If the joint proves to be economical, then this is a super system. One could cut the lengths of wood at home. Clever."*

"Juhuu!"

industrial design + product design

Design
Mohsen Jafari Malek
Tehran, Islamic Republic of Iran

Azad University of Tehran
Tehran, Islamic Republic of Iran

E-BOARD / IRONING BOARD
IRON BOARD

Ironing can be quite time-consuming. With the e-Board (easy-board), however, the user can iron clothes on the board much faster; once one side has been completed, the board can be flipped over on its central axis to iron the other side. This is the main advantage and saves time. The main board is connected to the base with a special part that makes it very easy to use. It is made of five parts, each of which can be adjusted separately to accommodate different types of clothing, like pants, shirts or skirts.

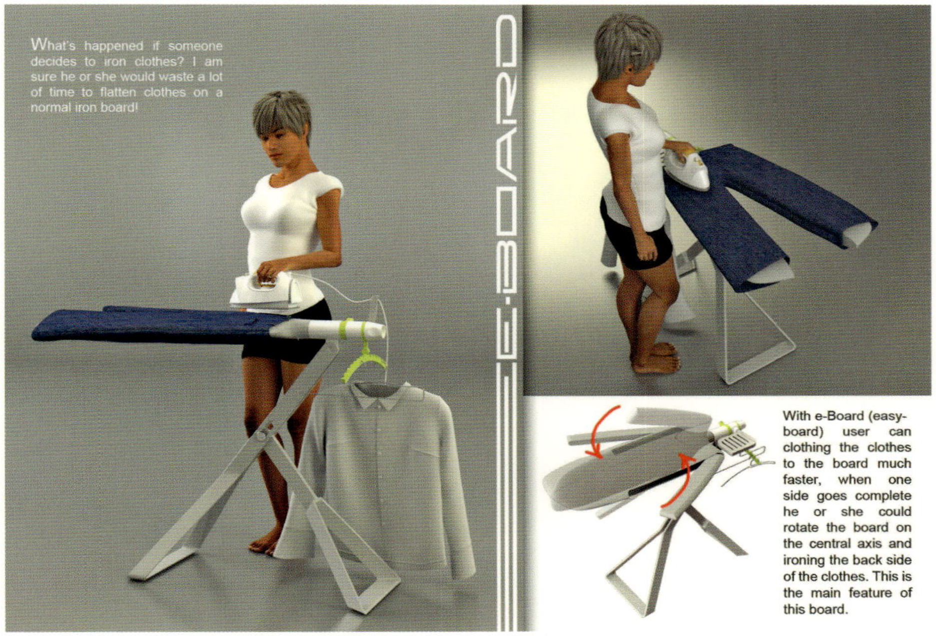

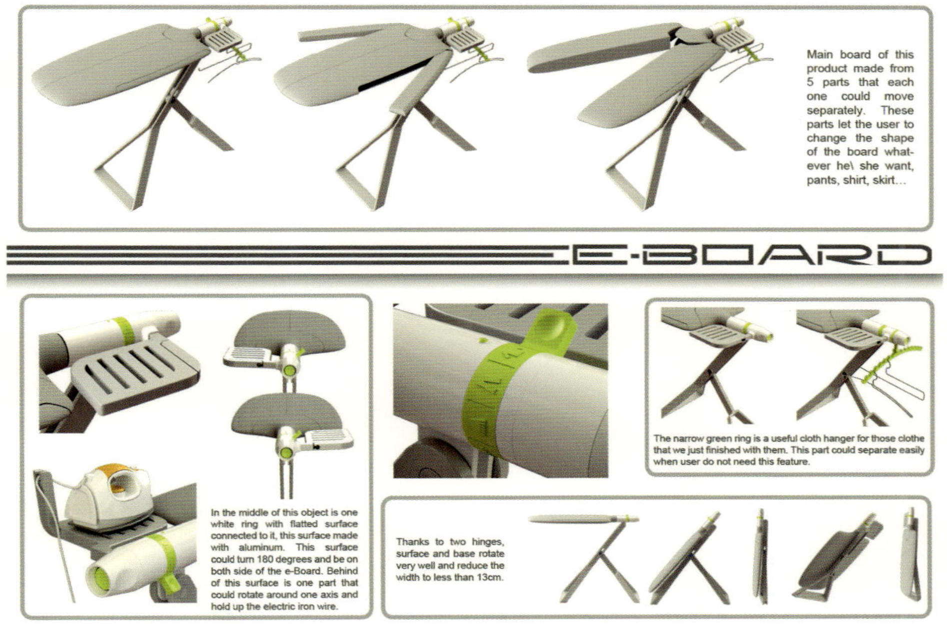

James Irvine: *"This is wild. If it works, it's fantastic. The Ferrari of ironing boards."*

"I have been pursuing my design career and participating in international competitions for a while. I always had the iF in mind though; to achieve winning one of these worlwide renown design awards was one of my goals. Winning this award makes me feel honored and proud and it gives me motivation for more hard work and passion in my career as a freelance designer."

Design
Sung-Young Eom
Cheongju, South Korea

Cheongju University
Cheongju, South Korea

ECO GLASSHOUSE
GLASSHOUSE

This greenhouse, in which crops can be grown in any season, includes solar panels, a snow-removal system and a heating system. The angle of the solar collectors can be adjusted to more effectively store the available solar power and avoid the use of fossil fuels, making greenhouse cultivation more economical than ever before.

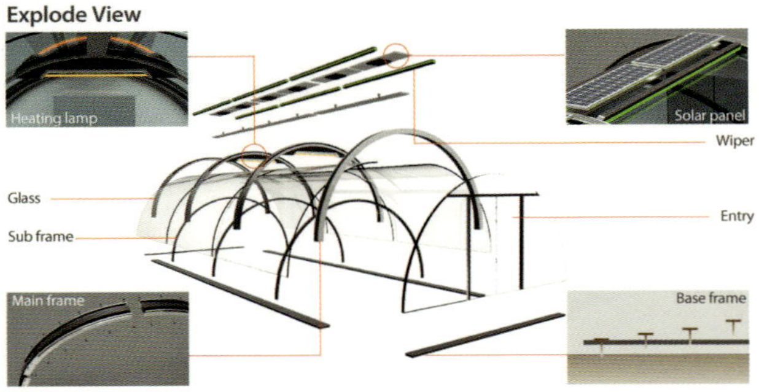

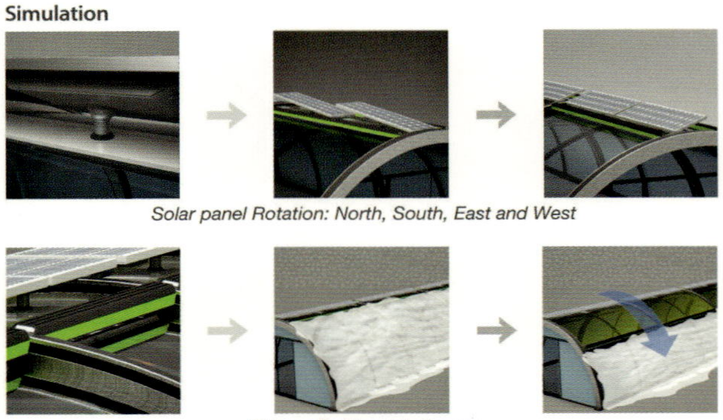

Martin Köttering: *"I am very happy that young designers are focusing on problems involving energy efficiency and food production to develop concepts that are socially meaningful. With the ECO glasshouse, Sung-Young Eom has succeeded in finding a sophisticated solution in terms of technology and design for a responsible way in which to reduce the amount of energy that greenhouses require."*

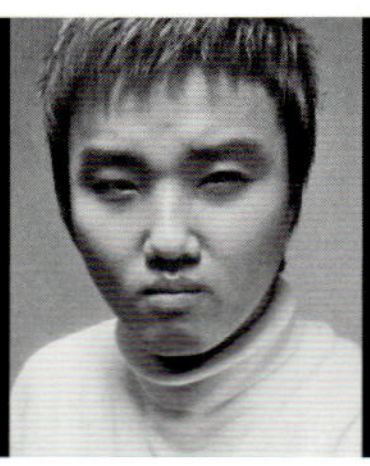

"I would like to thank my professors for giving me advice, and my friends for encouraging me during this project. Thank you very much."

industrial design + product design

Design

Frank Förster, Henrik Holkenbrink
Hannover, Germany

Fachhochschule Hannover
Hannover, Germany

ENDUR
STRETCHER FOR MOUNTAIN TERRAIN

The "Endur" stretcher is specially designed for use in remote mountain regions. Inspired by natives in the Himalayas, it can be used to carry patients to the nearest road to meet the ambulance. The most frequent emergencies involve pregnant women, which is why the Endur also allows the patient to sit upright. For greater comfort for the carriers, carrying straps can be used over level ground. On steep slopes, the front carrier can shoulder the stretcher. The stretcher narrows towards the front to enable the stretcher to be carried by 3 people on very narrow and difficult routes.

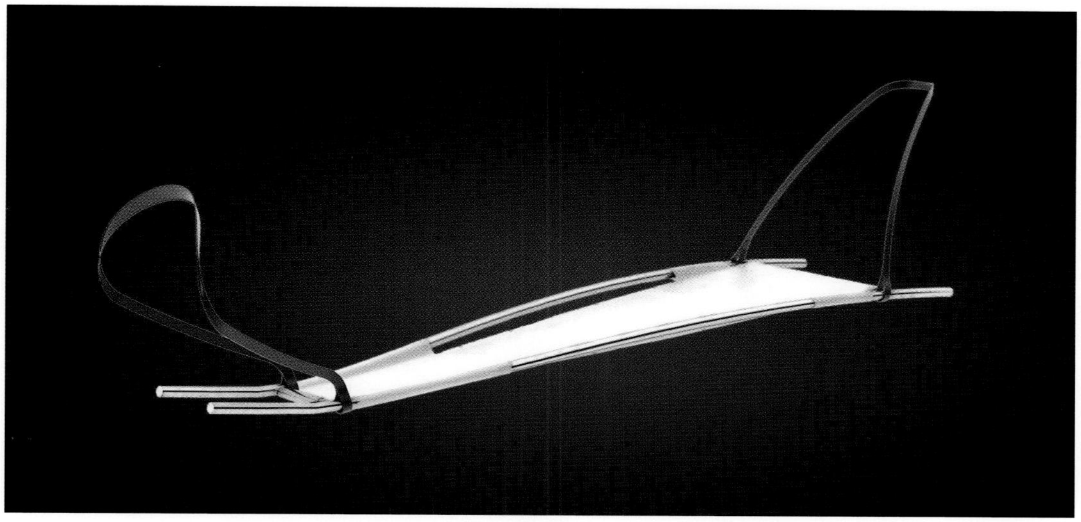

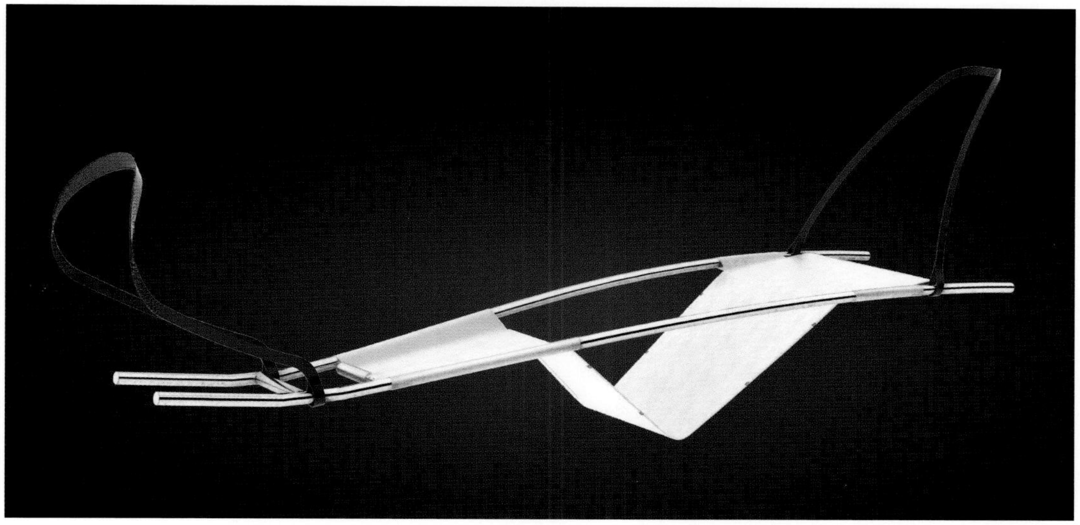

"Among so many good entries, we are really glad to be honored with an award."

industrial design + product design

Design
Kuo-Ting Chung
Zhongli City, Taoyuan County, Taiwan

Shih-Chien University
Taipei, Taiwan

E-STREAM
GUIDANCE SYSTEM

As a rule, power generating blocks and guide tiles for the visually impaired are both located at the entrances to mass rapid transit systems, but few people know how important they are. "E-Stream" combines power generating blocks with the guide tiles to transform people's steps into energy. When people step on the tiles, the tiles compress and this energy is fed into nearby batteries. This new product is also designed to make the general public more aware of guide tiles for the visually impaired.

"Thank you all, you guys make me who I am!"

industrial design + product design

Design
Cheng-Wei Wang, Jiuan Mau Tzeng,
Chi Wang (National Taipei University of Technology),
Chih-Wei Lai (National Taipei University of Technology)
Bade City, Taiwan

Ming Chi University of Technology
Bade City, Taiwan

FEAR NOT HAMMER

Hitting a tiny nail with a big hammer can be quite a challenge and sometimes it's your fingers that suffer. This is an even bigger problem for the elderly, whose vision is no longer 20:20, or for people who lack strength in their arms. Holding nails to hammer them in at a 45-degree angle is also a problem; just one slip and you could knock a hole in the wall. With this new type of hammer, you simply detach the end of the handle, insert the nail, hold it up against the wall and hammer the nail into place. You'll never hit your finger or thumb again!

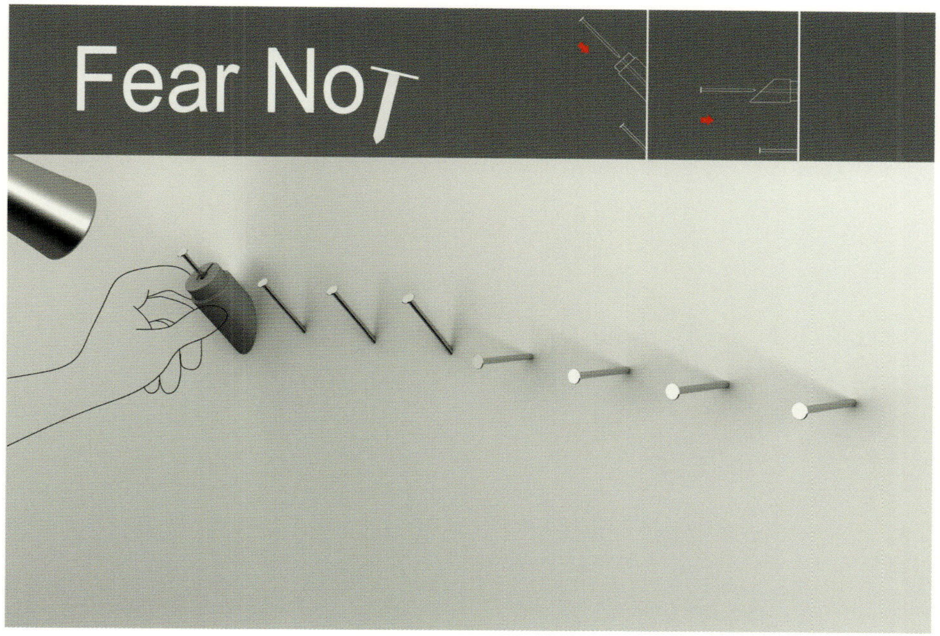

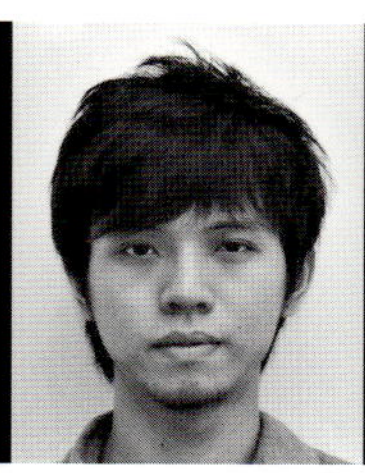

"May the odds be ever in your favor!"

industrial design + product design

Design
Ta-Chih Lin, Yi-Fan Hsieh
Taipei, Taiwan

Tatung University
Taipei, Taiwan

FINGER CONTACT
CONTACT LENSE PACKAGE

This is a new type of sterile packaging for contact lenses that keeps your fingers away from the lenses. It's like a protective shell. Now you can use the package to insert your contact lenses without worrying about getting any germs from your hands into your eyes. This is particularly convenient and offers optimum hygiene when traveling or for use during outdoor activities.

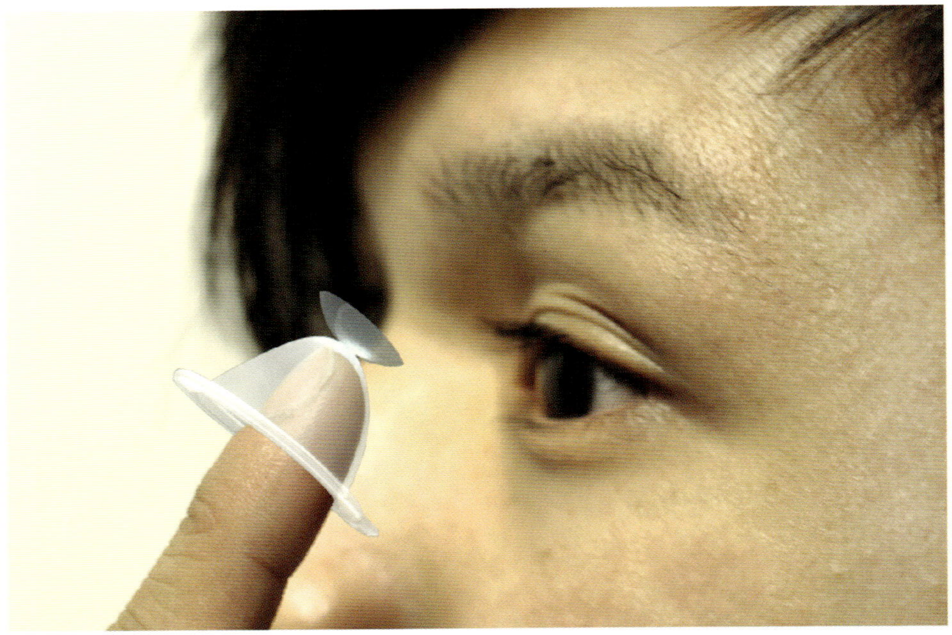

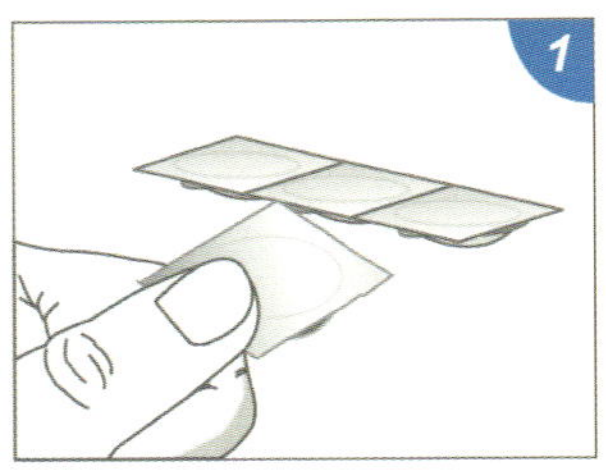

PICK OUR PRODUCT

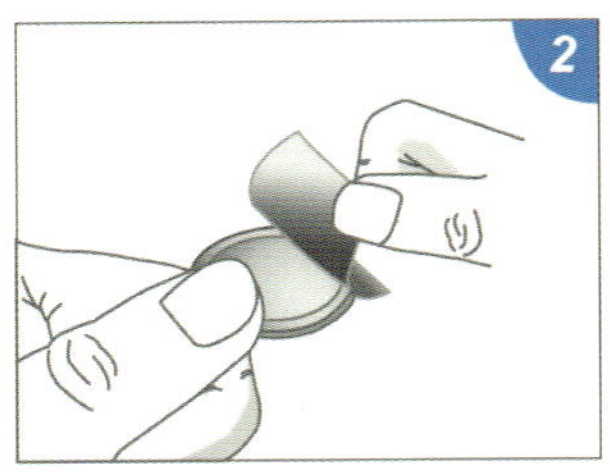

TEAR THE PACKAGE

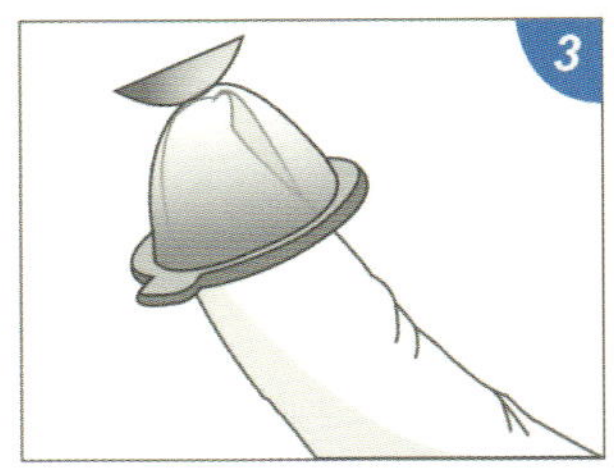

INSERT YOUR FINGER

"We yelled 'Wow!'. Words cannot do justice to how we feel. To us, the award is a significant milestone that pushes us to continue seeking out great designs. We hope to be back soon!"

industrial design + product design

Design
Hu Yuan
Guangzhou, China

South China University of Technology
Guangzhou, China

FUNNY SHADOW
TOY

Some kids like to draw patterns and pictures on the wall or the ground when they're playing. This has a negative impact on the environment and is often difficult to clean up. This product is a portable toy, which can project images, e.g. the graphic patterns of different games that children like to play. Children can control the game by turning the "ring". The patterns can be downloaded from the internet.

Adrienne Viljoen: *"The entry impresses by using technology to give children access to a variety of well known traditional games as illustrated as well as the potential to create new games. It impresses as it encourages children to rethink possibilities and to use and adapt according to experience. It is attractive and well presented. It is usable in many situations and there is no safety risk."*

"When I heard the good news, I was so excited and thought – a miracle! Thank you for this award! I feel very proud. I believe that the iF concept design award encourages me to create better design."

industrial design + product design

Design

Fan Yang, Yiping Zhou, Miaomiao Lin, Yifan Zhou
Dalian, China

Dalian Polytechnic University
School of Art and Design
Dalian, China

GESTURE TRANSLATER APPLICATION

The concept of this design is to enhance the communication skills of anyone with perception disabilities (such as speech or hearing impairment). It is important to realize what difficulties they face when they try to communicate with anyone from the general public. Because the majority of the general public is not trained in sign language, this mobile application has been designed to translate sign language into verbal communication and to translate sound into texts to enable communication in both directions.

"We are so excited to receive the award! We prepared two months for the iF competition, we aimed to help people solving problems in their daily life and make things universal, and the idea of enhancing the communication abilities of anyone with perception disabilities (such as speech or hearing impairment). Finally, a mobile application was popped up, the mobile is for communication, and we would love to expand the function. The prize means a lot for us, and it is also an encouragement, we will keep designing, because we love design!"

industrial design + product design

Design
Szu-Yu Liu, Ming-Yu Tseng
Taipei, Taiwan

National Taiwan University of Science and Technology
Taipei, Taiwan

GO DUTCH
BILL

Sometimes things can get difficult when a group of people dine together at a restaurant, but then want to pay separately. Either the server has to split the bill or the people in the group figure it out for themselves. "Go Dutch" is a concept that solves this math problem in a clever way. The bill itself is divided into parts along perforated lines so that each customer can tear off their own portion and pay their own bill.

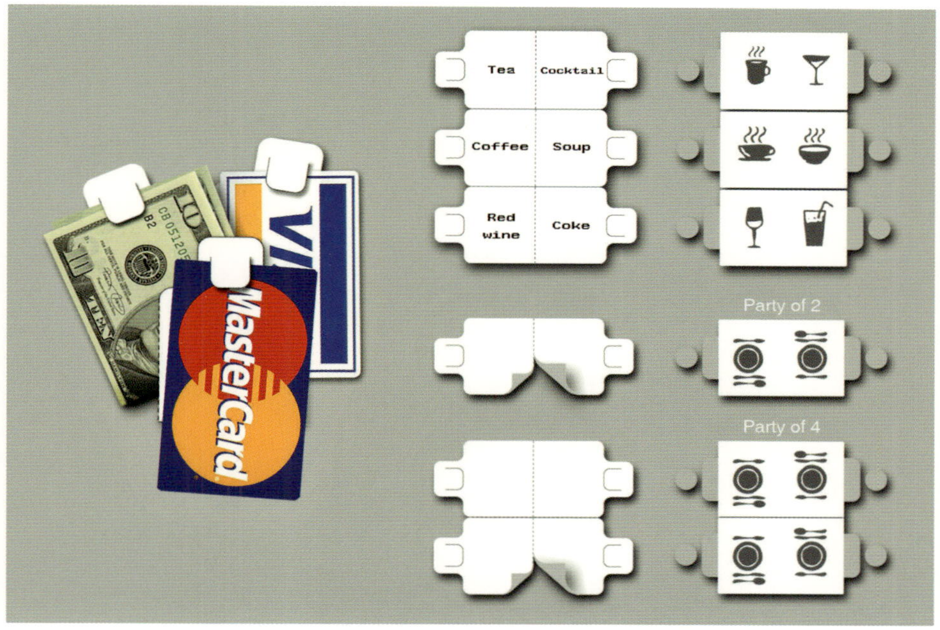

Julian Lindley: *"A refreshingly simple observation continued into a practical and fun solution. The idea has been quickly communicated. A good example of how a designer can see new possibilities in the mundane by observatiion."*

"As designers, we are always inspired by daily objects and try to turn them into design opportunities. As designers, we are privileged to process the knowledge and techniques to visualize our thinking. As designers, we are happy to participate in the iF concept design award and show our ideas to the world. As designers, we're extremely excited and honored to receive this meaningful prize. Thank you!"

industrial design + product design

Design
Matthias Lauche
Hannover, Germany

Fachhochschule Hannover
Hannover, Germany

GRIFF
HELPER FOR EVERY DAY THINGS

"GRIFF" is a product developed for children, seniors and disabled people who have limited mobility in their hands and fingers. This ergonomic tool makes it easier to grasp and pick up such things as pens, cutlery, cooking utensils, wine glasses and more. The handy slot makes it easy to handle: simply slide the things in. The flexible material accommodates all kinds of different objects of different sizes and shapes. Smart, simple and colorful little helpers.

Peter Kuchnicki: *"Simple and excellent example of universal design. Many people could benefit from its combination of greater convenience and safety. Sometimes "simple" is the essence of a solution."*

"I was surprised to hear that I had made the shortlist, considering the unbelievable number of entries that were submitted this year. To be chosen as one of the best 100 of 10,665 is incredible. Once the news had sunk in, I was very, very happy."

industrial design + product design

Design
Fangtian Ying, Guanyun Wang,
Chengyuan Wei, Ye Tao, Lei Shi
Hangzhou, China

Zhejiang University
Hangzhou, China

HAPPY-SCOOTER
SCOOTER

Happy-Scooter turns a fixed mobility tool into a changeable and interesting game. Happy-Scooter employs a simple structure and moveable handles so that there are two ways to use it – standing or sitting. Its structural frame and handles are made of wood and bamboo, which is 'soft' to the touch and easy to handle. The scooter components can also be easily assembled and replaced. Happy-Scooter provides for a refreshingly new riding experience and a creative relationship between the rider and the scooter.

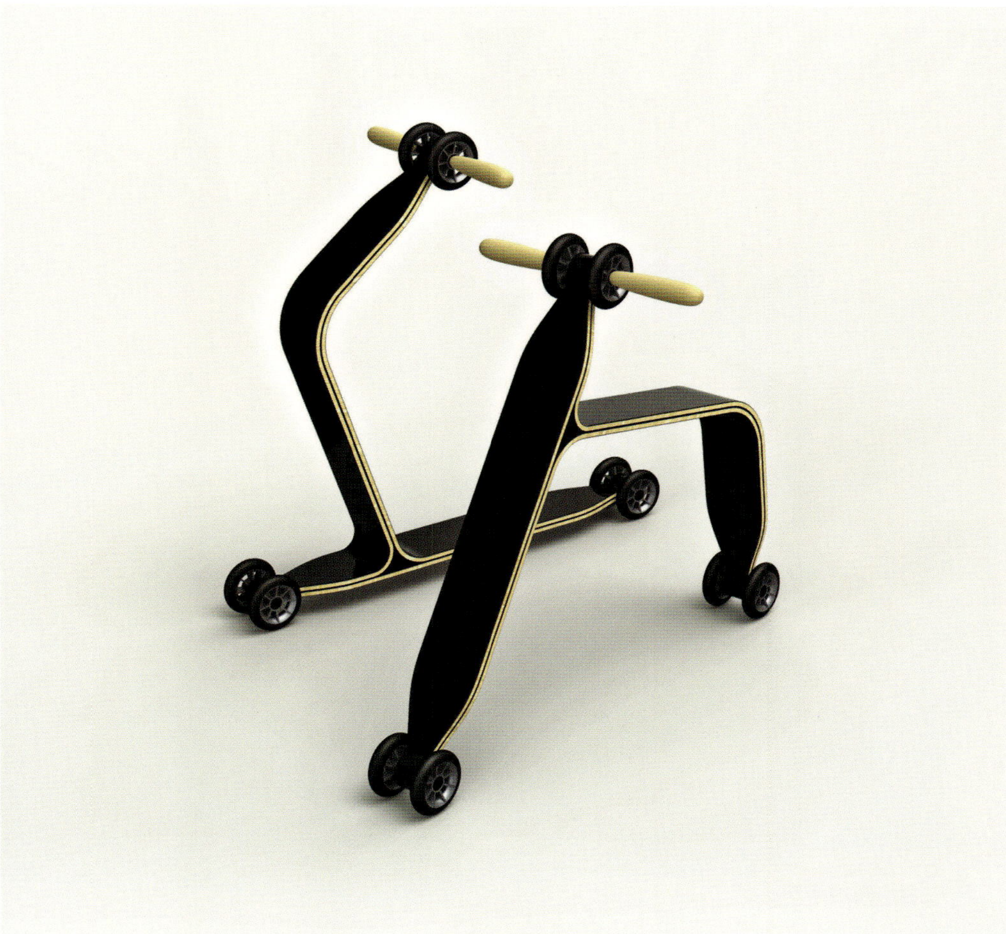

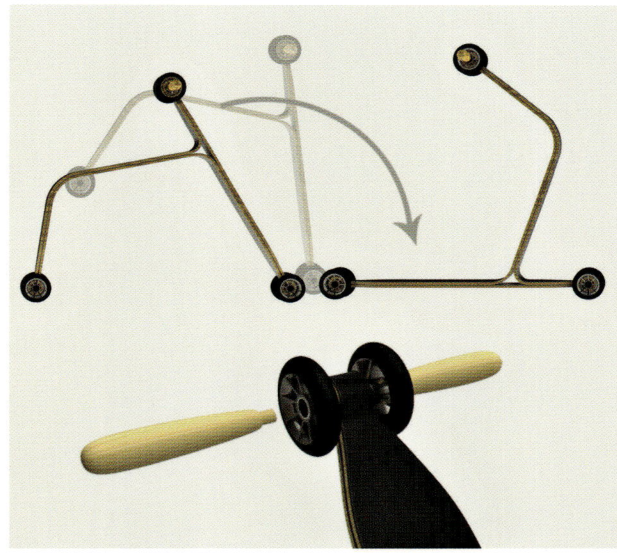

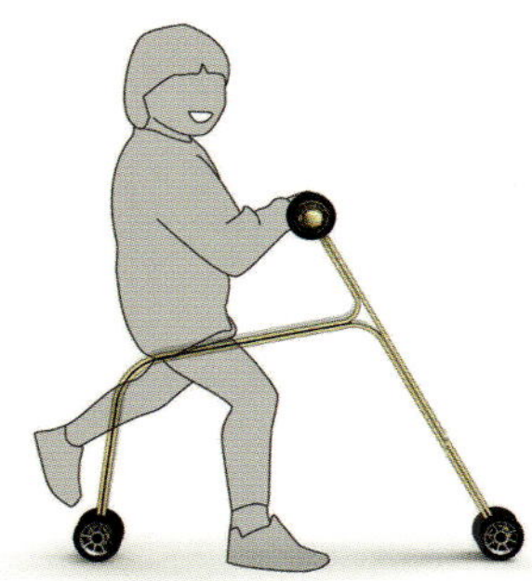

"While exploring what is understood by and the need for originality, morality, emotion and aesthetics in this world, we did research on the aged, children and the handicapped. It is necessary to carry on and develop more designs for the future."

industrial design + product design

Design
Fangtian Ying, Fan Zhang, Guanyun Wang,
Yue Wu, Xuanxing Yang
Hangzhou, China

Zhejiang University
Hangzhou, China

INJECTION ASSISTANT
MEDICAL DEVICES

This product is a non-contact vein illumination device. It can make superficial veins appear more noticeably different than the surrounding tissue because the hemoglobin in the blood absorbs the infrared light. It can be used to locate veins to take blood samples or to start IVs more easily and to significantly reduce the number of needle insertions that miss the vein.

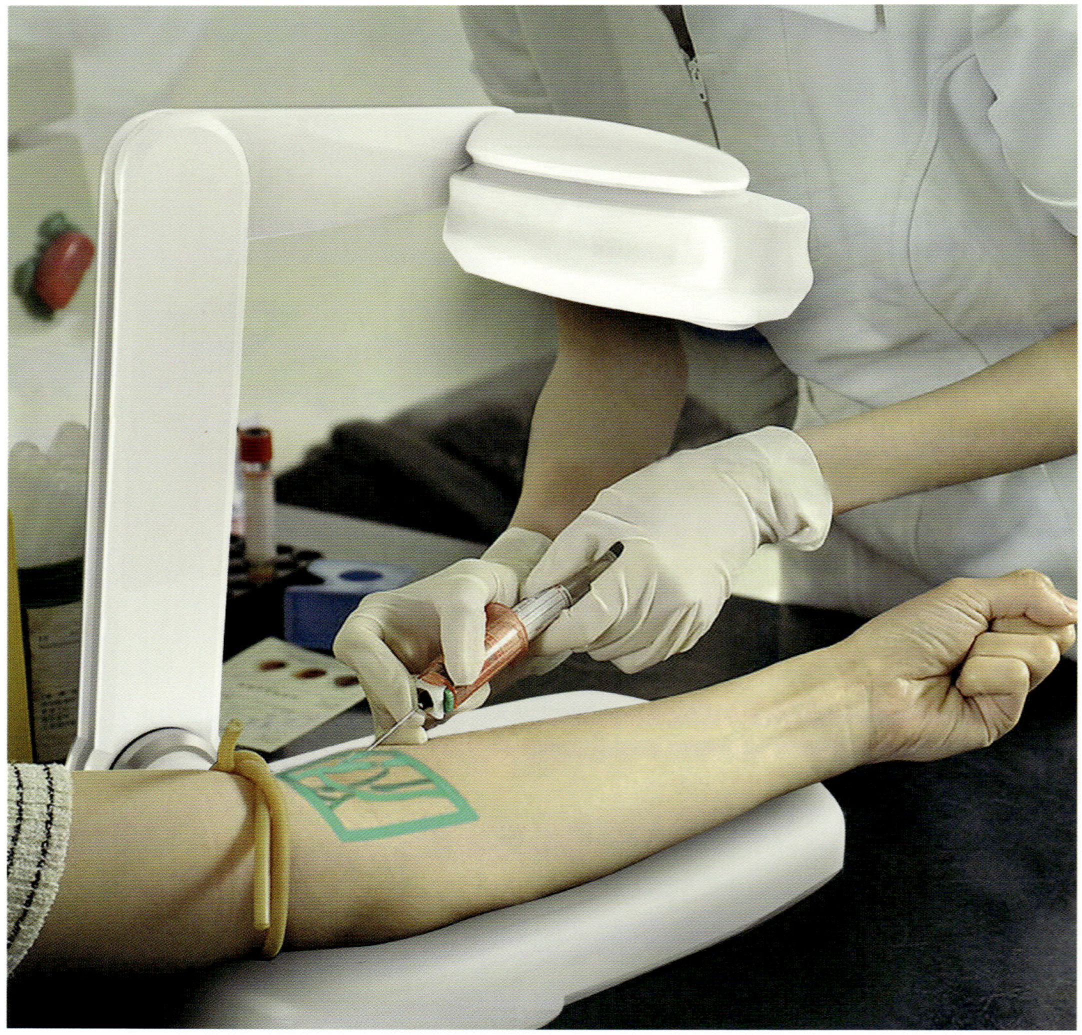

William Wu: *"Smart idea that utilizes technology so well. No more unnecessary pain."*

"While exploring what is understood by and the need for originality, morality, emotion and aesthetics in this world, we did research on the aged, children and the handicapped. It is necessary to carry on and develop more designs for the future."

industrial design + product design

Design
Simon Fredriksson
Umeå, Sweden

Umeå Institute of Design
Umeå, Sweden

NIOX MINE
MEDICAL DEVICE

"Niox mine" is designed as a complete system to provide the necessary equipment and guidance for patients diagnosed with asthma. "Niox mine" allows the user to monitor the level of inflammation by measuring the nitric-oxide level in the airways as an indicator of how strong the inflammation is. This information makes it possible for the user to adjust medication dosage as needed to prevent over or under-medicating and helps asthmatics to enjoy a more problem-free life.

Boudewjin Soetens: *"A very well detailed and clearly presented product design that serves a specific user need."*

"Receiving an iF concept design award is a great honor and a proof that what I do in design is appreciated by others. The shear number of participants this year makes it even better because then you know that you have been judged against a large number of great designs. I´m proud of the award and glad that I can help keep proving that Umeå Institute of Design is a world class design school."

industrial design + product design

Design
Ilmo Ahn, Donghan Seo, Taehan Kim
Seoul, South Korea

Konkuk University
Seoul, South Korea

INVERTED UMBRELLA
UMBRELLA

This innovative umbrella has been designed to fold up in reverse, so that the dry inner side of the umbrella is on the outside. Conventional umbrellas have the wet side on the outside with the unpleasant side effect that they often get clothes and other things wet after they have been closed.

"I just hope that my designs can be helpful to many people. Winning this award is an important step along the way."

industrial design + product design

Design
Di Fang, Xuefei Liu, Yina Liu, Pei Liao
Dalian, China

Dalian Nationalities University
Dalian, China

IRING
MOBILE PHONE HOLDER

The touch-screen cell phones currently available on the market now have an increasing number of functions. They can be used to read documents, view photos or watch videos. However, holding onto a phone for an extended period of time can make your hands tired and you might drop it. "iring" is a specially designed product that solves this problem. It is equipped with a suction cup to grip the back of the cell phone and a ring that slips over your finger so that the phone won't slip out of your hand.

Angela Carvalho: *"A small detail that enhances the usability of mobile phones, making them easier and safer to handle."*

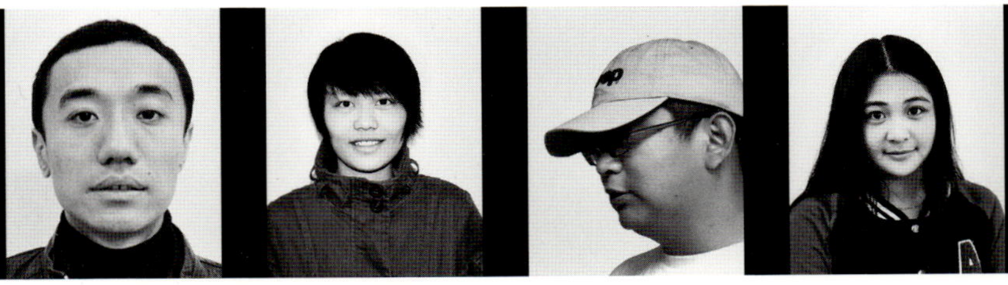

"We are concerned about enhancing life and helping people through design. Our work has been recognized by the judging panel, we are very excited and happy."

industrial design + product design

Design
Xu Jiang, Wang Zihao, Feng Lu, Liang Qing, Zhao Yingzong
Wuxi, China

Jiangnan University
School of Design
Wuxi, China

JUICE BOTTLE
JUICER

This smart juicer lets you enjoy fresh juice any time anywhere. All you need is the bottle and an orange and you can prepare your juice wherever you happen to be – a great alternative to having to buy artificial canned beverages.

"We are so excited to get the iF concept design award, a world-class award that is recognized internationally. Thank you, iF, for giving us a chance to demonstrate our creativity and make our dreams come true. Thanks also to our university. In the future, we will continue to forge ahead along the path of design innovation."

industrial design + product design

Design
Damian Schneider, Matthias Scherzinger
Zürich, Switzerland

Hochschule für Gestaltung und Kunst – HGK FHNW
Aarau, Switzerland

KANDO
COMBINATION BACKPACK / TENT

"KANDO" is a new outdoor concept that combines an alpine backpack with a one-man tent in a single product. Thanks to the ideal incorporation of the individual parts and the choice of materials, a product was created with a special focus placed on volume reduction, weight optimization and easy handling. The tent is based on half-arch design held in place by guy-wires. The middle part of the backpack can be unzipped so that the user can choose between using the full touring pack (+33L) or the compact daypack (+18L).

"We are delighted to have won the iF concept design award and hope that it will open more doors for us, enabling us to actually get KANDO and other projects from the design stage to the market."

Design
Ziran Zhao
Beijing, China

Tsinghua University
Beijing, China

LIFE PASSAGEWAY
UNDERPASS

"Life Passageway" is a small, triangular tunnel laid under expressways so that the small animals living in that area can travel through them without being hit by vehicles. According the estimation of HSUS, there are more than 1 million animals killed in traffic accidents every day. "Life Passageway" does not only reduce road kill, it also makes driving on the expressways safer. In addition, the triangle shape makes the structure more stable. Individual modules can be easily combined to accommodate different road widths.

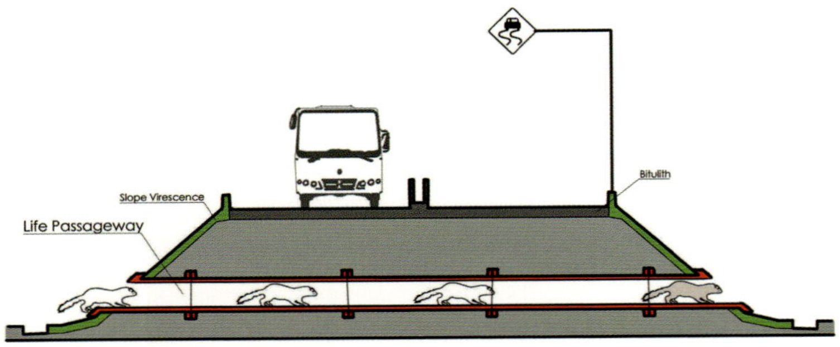

"I am really grateful that the jury selected me as one of the winners of the iF concept design award. For me, it was the most exciting experience that I have ever had, and I think that this award is the highest recognition of my works and study at university. Thank you."

industrial design + product design

Design
Po-Yen Yeh
Taipei, Taiwan

National Chang Gung University
Tao Yuan, Taiwan

LIGHTWAVE
LIGHTING

"Lightwave" is a mood light designed to create a pleasant atmosphere in a public space, while at the same time acting as a WIFI hotspot. Lightwave uses the latest advancements in flexible solar panels to power the LED lights as well as the WIFI station, all of which is concealed within the two curved metallic enclosures to recreate the rhythmic, uniform movements of a school of stingrays for a life-like marine visual in the overall design.

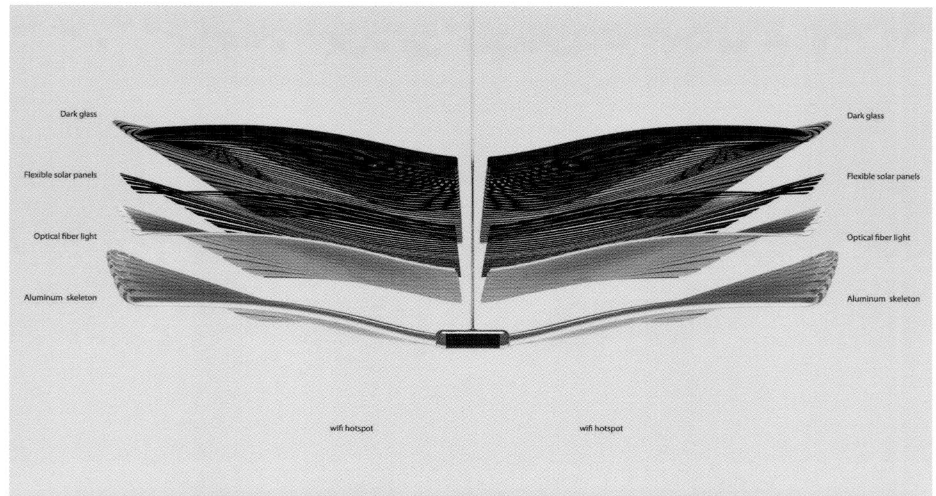

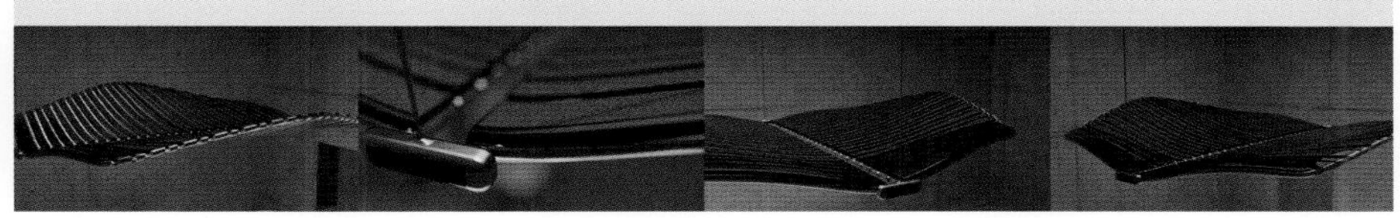

Simone de Waart: *"Beautiful object that cleverly combines different layers of light and energy."*

"The many beautiful marine habitats in and around Taiwan served as the source of my inspiration. I am honored to have my design entry selected by this year's iF concept design award judges, and would like to share this joy with everyone involved. Thank you for your guidance and unreserved support during this endeavour."

industrial design + product design

Design

Xu Zhang, Yilin Wang, Yichuan Lu, Kaiyan Tian
Xi'an, China

Xi'an University of Science & Technology
Xi'an, China

LIVE SOCKET
SOCKET

When you use electrical appliances, you always have to plug them in and then unplug them again, which can be annoying. They should also be well out of the reach of children to avoid accidents. This product design solves these problems and provides for more convenient use for everyone.

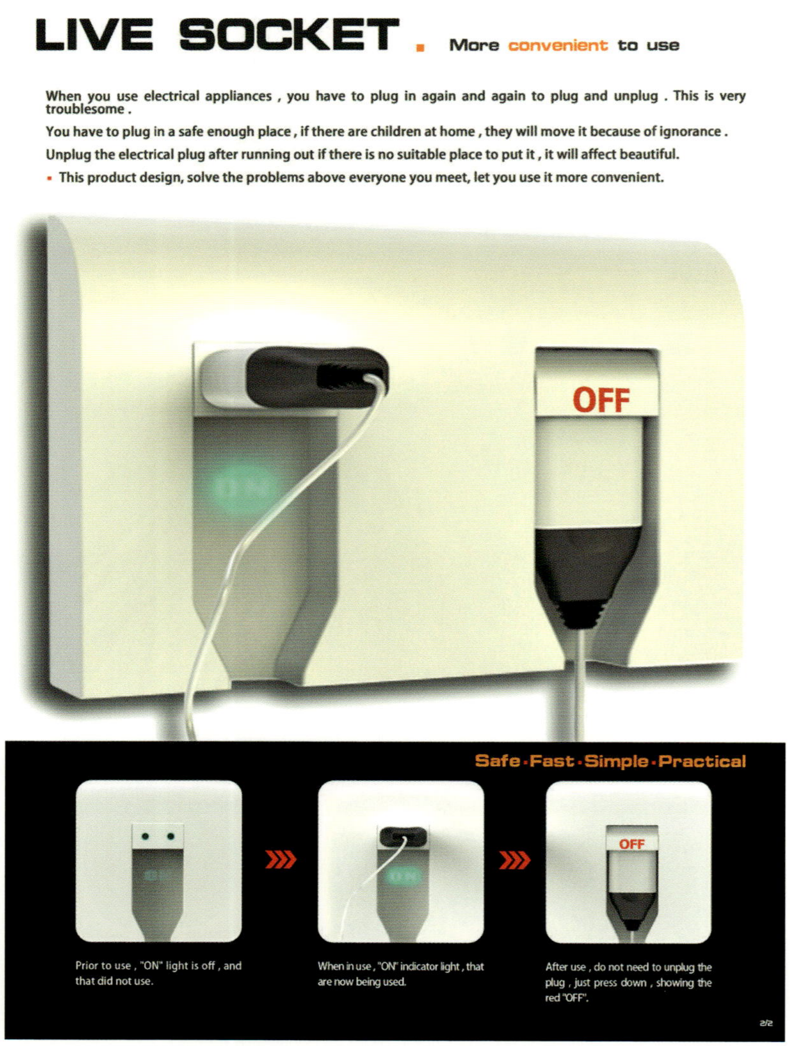

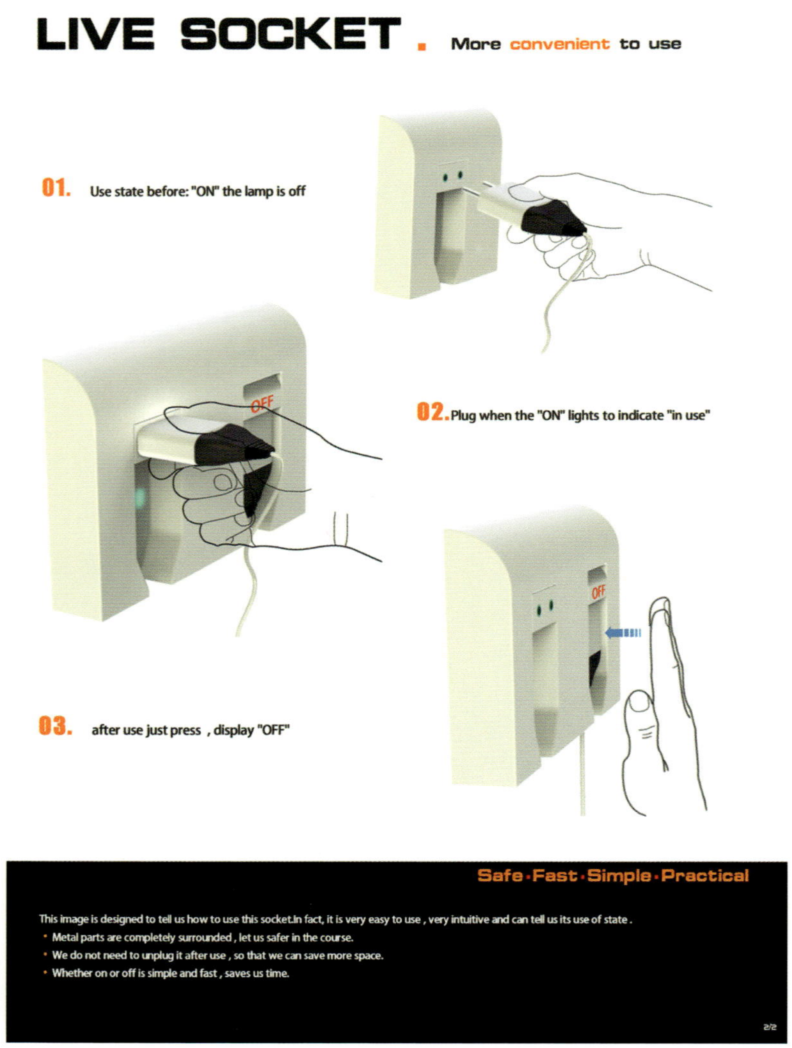

"We are really pleased to receive this award. iF gave us a good opportunity to show what we can do. But I think that design should be an innovative process, I will continue to improve it."

industrial design + product design

Design
Yasaman Sheri
Montreal, Canada

Concordia University
Montreal, Canada

LOTA AGUA
SHOWER

"Lota Agua" is a low-cost folding shower for use in areas where there is no running water available. The user simply stands on a container filled with water. The water pressure required is created by the user's body weight – a valve on the shower head controls the water flow. The shower is very easy to set up. The water used for showering can then be directly recycled for watering gardens and plants.

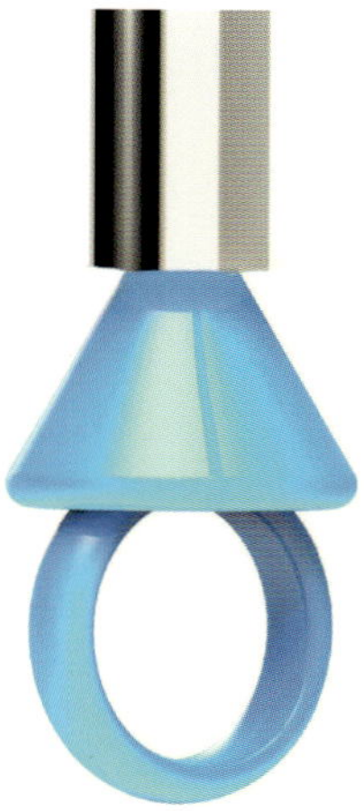

William Wu: *"Good way to help people in undeveloped areas and enhance everyday life. This is the power of design."*

"I am absolutely honored to receive this award as a Canadian designer and hope for more countries around the world to create award-winning designs recognized by iF."

industrial design + product design

Design
Di Fang, Xuefei Liu, Long Lu, Jinghua Guo
Dalian, China

Dalian Nationalities University
Dalian, China

MAGICAL MANIFIER
ELECTRONIC MAGNIFIER

This is a magnifying glass that can change the magnification of the words by adjusting the thickness of the glass. This makes it more convenient to use than other magnifying glasses and it looks good, too.

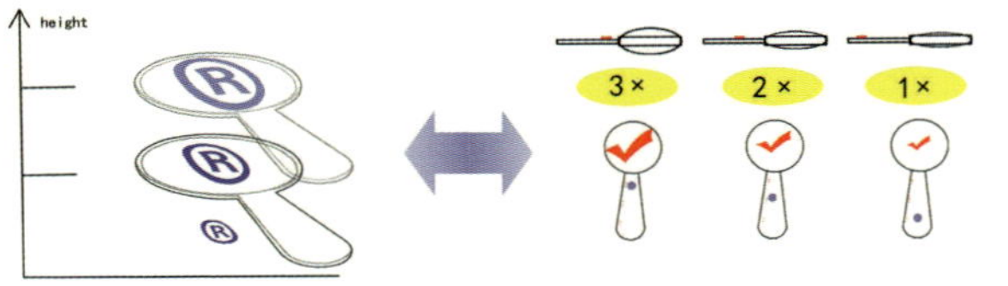

adjust the place to change the size of the words

Push the blue spanner can adjust the thicknner of lens , so we can change the size of seeing words. This scheme avoid trouble when moving the magnifier.

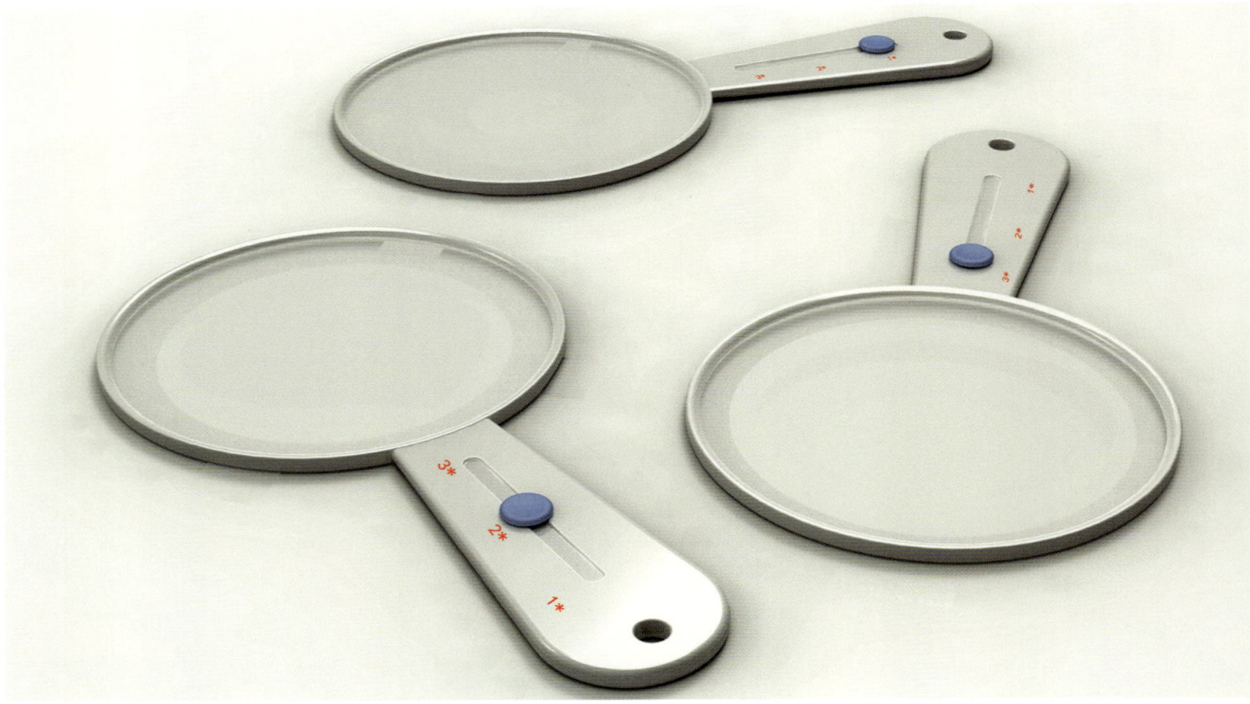

"When we heard the news about winning the award, we got very excited. In the future, we will continue our efforts to design better things."

industrial design + product design

Design
Jichang Kang, Jangsoo Kim (Bucheon University),
Jiyun Lim (Joongbu University)
Seoul, South Korea

Korea Polytechnic University
Gyeonggi-do, South Korea

NONSLIPER
CRUTCH

The "Nonslipper" is a crutch designed to provide adequate support to the user even on slippery surfaces. It can be easily adjusted for people of different heights by simply rotating the handle. The user can also adjust the height appropriately for additional support when seated, for example, and fold it up easily for storage when it is not in use. The arched rather than straight shape also minimizes the loss of kinetic energy and reduces impact.

"We as a team are very honored and happy to win this iF concept design award, despite our lack of experience. We look forward to enjoying this honor along with our parents, who have trusted and supported us, the professor who taught and encouraged us a lot, and our friends."

industrial design + product design

Design

Eun ah Kim, Su hyun Yoo, Ju hyeon Lee
Seoul, South Korea

Kookmin University
Seoul, South Korea

OASIS
WATER BOTTLE

Natural disasters are a common occurrence all over the world and the supply of clean drinking water is often affected. While the average person needs half a gallon of water a day to survive, people can't always carry a sufficient amount of drinking water with them. The bottles are too big and too heavy. The "Oasis" is a flexible bottle made of the latex used for condoms. You can only see the top part of the bottle, so it's small, light and easy to carry. Inside, there is a flexible latex ring. In an emergency, you can quickly fill the bottle with up to 20 liters of water from a reliable source.

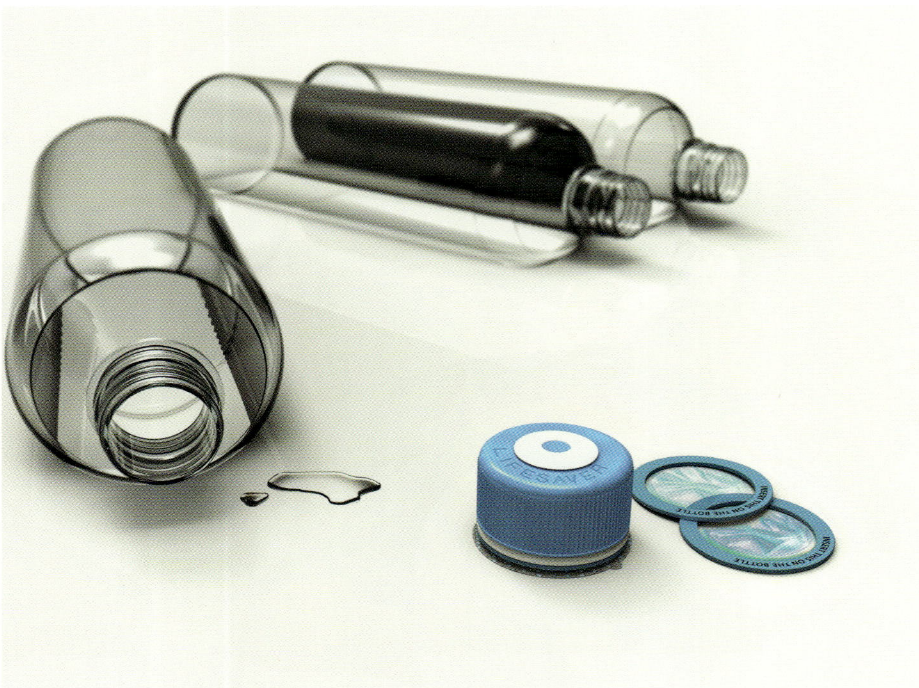

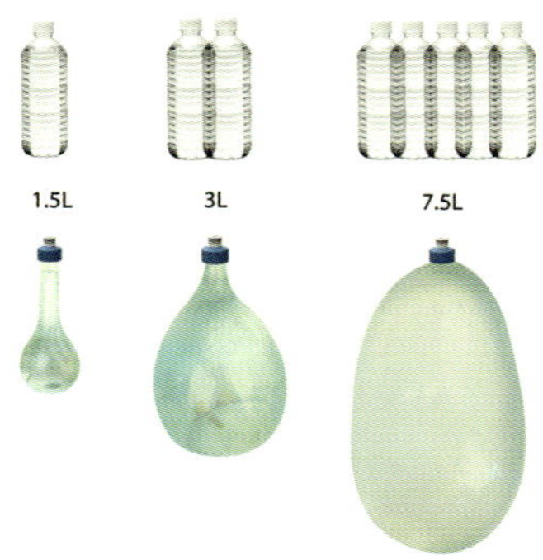

William To: *"Light and handy solution designed to save more water during a crisis."*

"Wow! We can't believe that we won! Thank you for giving us a great prize. We learned a lot while we were doing this work. We were delighted to have a chance to show our concept on the iF concept design award platform. Thank you."

industrial design + product design

Design
Leon Chen
Chiayi County, Taiwan

Nan Hua University
Chiayi County, Taiwan

PIEZOELECTRIC COURT
ENERGY USAGE

Lighting up a sports field or a court at night consumes a great deal of electricity, so this concept employs PZT and the energy produced by the athletes themselves to save energy and still light up the area. This lighting element uses the pressure of every step taken by the athletes when they move or hit the ball, no matter what time of day, and converts this pressure into an electric charge to be stored in a battery. This is able to provide enough power to light up the sports field or court at night.

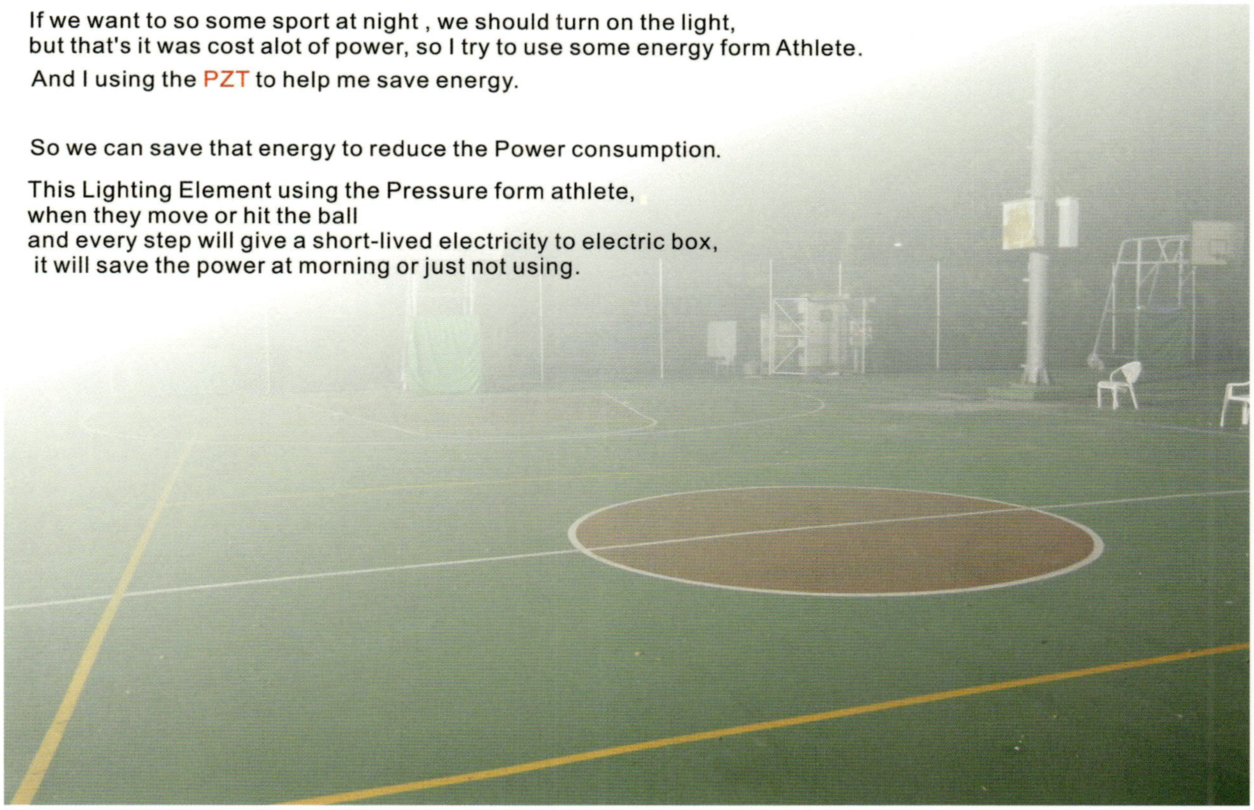

Ake Rudolf: *"The technologies to generate energy kinetically that this concept is based on are already in use so that it would be easy to transform this design into an actual product, even though production costs are likely to be relatively high at the current time. Another interesting aspect is the instinctive communication of the use of regenerative energies to the younger generations in particular. The concept also offers a variety of different areas of application for the use of the energy generated."*

industrial design + product design

Design

Ming-chieh Yeh
Taipei, Taiwan

National Taiwan University of Science and Technology
Taipei, Taiwan

PILL JAR FOR PD
PILL DISPENSER

Many patients suffering from such illnesses as Parkinson's disease (PD) have great trouble in controlling their shaking extremities, which can make actions that are normally easy very difficult. This includes getting the medication they need out of the bottle. "Pill Jar" helps the sufferers of PD (or the elderly) to get the pills out of the bottle easily, even when their hands are shaking. They get one pill out at a time, simply by pressing and releasing the top. In addition, the main part is reusable and, thanks to its silicone ring, it can be adapted to fit on differently sized pill jars. Using the "Pill Jar" requires no detailed instructions.

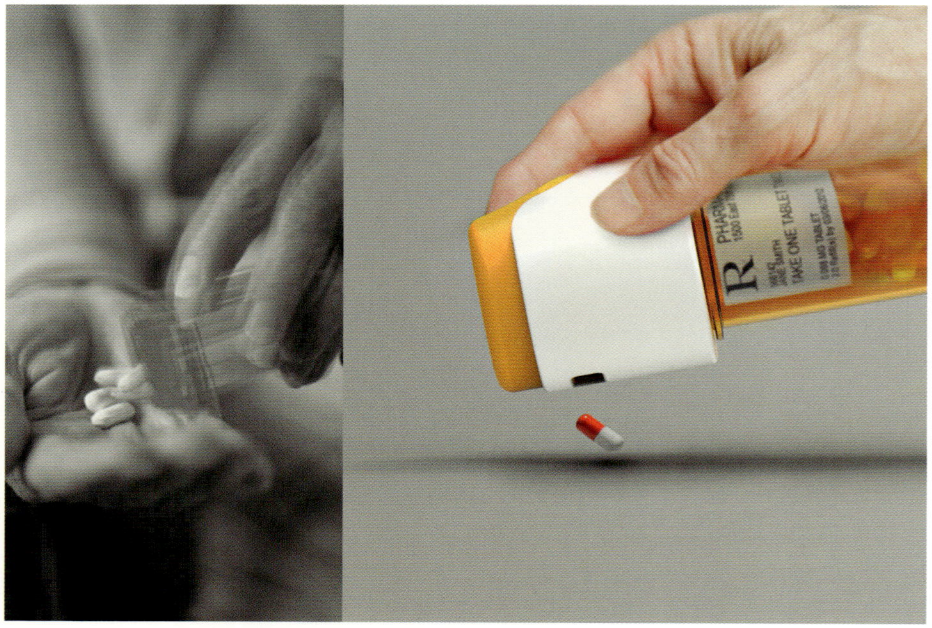

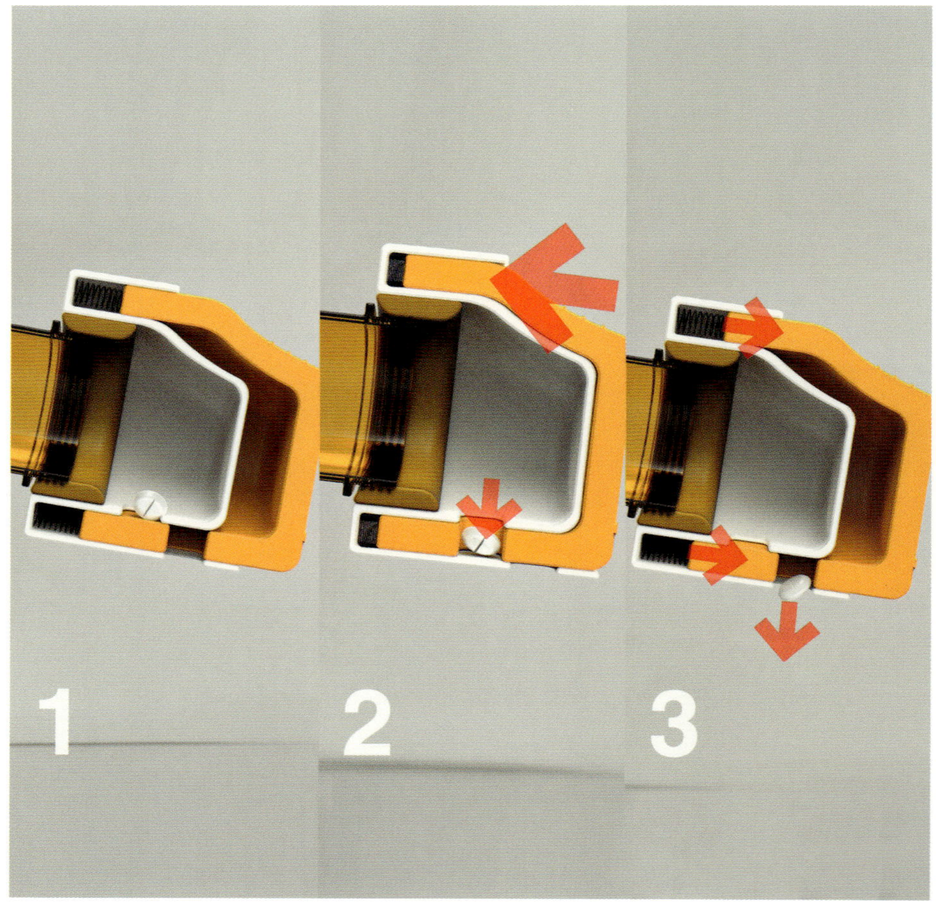

"When I got the good news, I shared it with my family and friends immediately. Undoubtedly, it's a huge encouragement to me and supports me in what I do. This special award gives me a chance to prove the quality of my design."

industrial design + product design

Design

Jun Ho Kim
Gyeonggi-Do, South Korea

Hyupsung University
Gyeonggi-Do, South Korea

POD LID

POD LID

Pot lids usually play a very limited role in the kitchen. When turned upside down, however, this pot lid is designed to serve as a container for the ingredients that have been prepared to be added to the pot. Divided into four sections, it is also easy to add these ingredients separately one at a time. The lip of the pot lid is narrow and thin to provide for greater control and to prevent the ingredients from falling off.

"I am very glad to get such a wonderful award. Everyday cooking in the kitchen inspired me to design this pot lid to be useful for more than just covering a pot. Thank you so much for this award; it has given me greater confidence in my ability to design."

industrial design + product design

Design
Julian Rathmann
Adendorf, Germany

Hochschule Coburg
Coburg, Germany

POLARWOLF
SLEDGEHOCKEY SLEDGE

SledgeHockey is ice hockey for athletes with lower-body impairment. The "polarwolf" is a highly functional, adjustable hockey sledge. Due to its modular design, it is suitable for most types of athletes and can be repaired easily and quickly in case of damage. In addition to its adjustable length, the "polarwolf" allows the hockey players to adapt the sledge to their own specific needs. It is also a serious, modern piece of sports equipment that does not stigmatize the user. The systematic use of innovative materials and manufacturing processes permits affordably priced production, even in small batches.

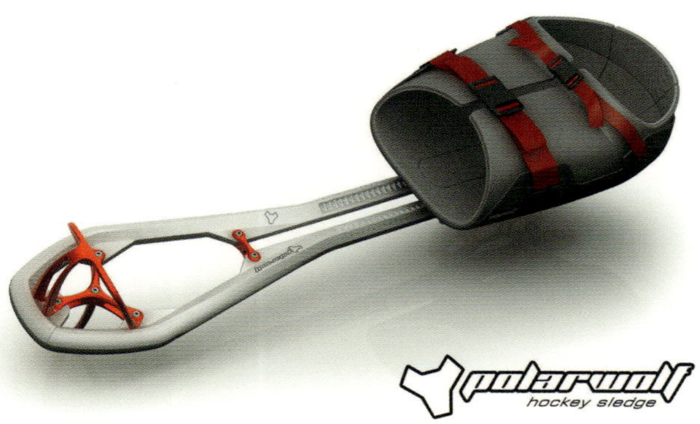

"I am very happy that my thesis was chosen to receive an iF concept design award. I am proud that it won when there were so many other participating entries this year. Winning this award is a great way to finish off my studies."

industrial design + product design

Design

Runiu Fu, Luting Liu
Tianjin, China

Tianjin Academy of Fine Arts
Tianjin, China

PRECIOUS WASTEWATER
WATERSAVING TOILET DEVICE

People produce a large amount of wastewater every day and the water just goes down the drain. However, domestic wastewater can also be recycled to make a valuable contribution towards environmental protection. This special toilet bowl features a water tank that is used to collect wastewater, like the water used for washing vegetables, showering or rainwater. First, the water tank is used to collect the wastewater. Then it is installed on the toilet bowl and the wastewater is used to flush the toilet.

Peter Kuchnicki: *"Clean water is in short supply everywhere. Every method should be explored to extend this limited resource. This design offers a non-mechanical way to capture wash-water and re-use it in the septic system. Nice."*

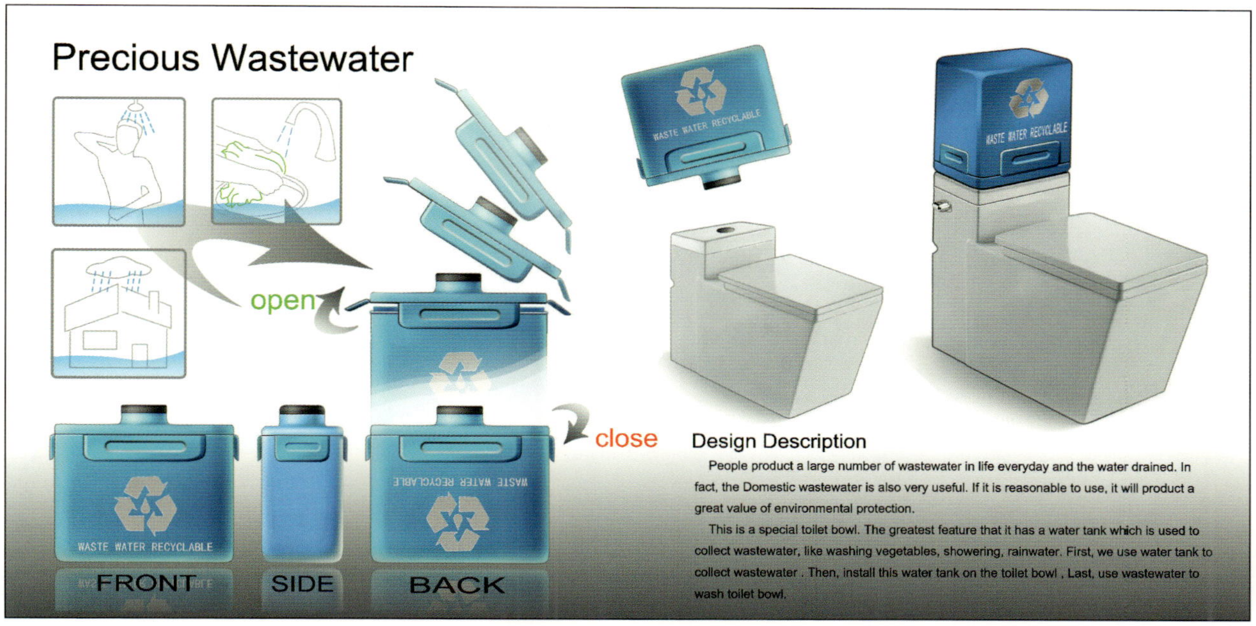

"We just love design for peoples every day life.
So we are so happy to win this competition."

Design
Ernst Koning
Amsterdam, Netherlands

Hogeschool Utrecht
Utrecht, Netherlands

REFINERY
LIGHTING

The "Refinery" is a source of light as well as a connection to the power grid and the internet. It can be placed above a dinner table, a conference table or a desk. It reflects simplicity, beauty and functionality. Its design was inspired by an oil refinery. With the "Refinery", industry enters our living room, office or studio in an adapted form to encourage an open-minded, involved and optimistic look into the future. It's made of stainless steel or aluminum and PU rubber and uses energy-efficient light bulbs, such as CFLs or LEDs. The user can turn the tubed arms, providing light, power and an internet connection, into any desirable position.

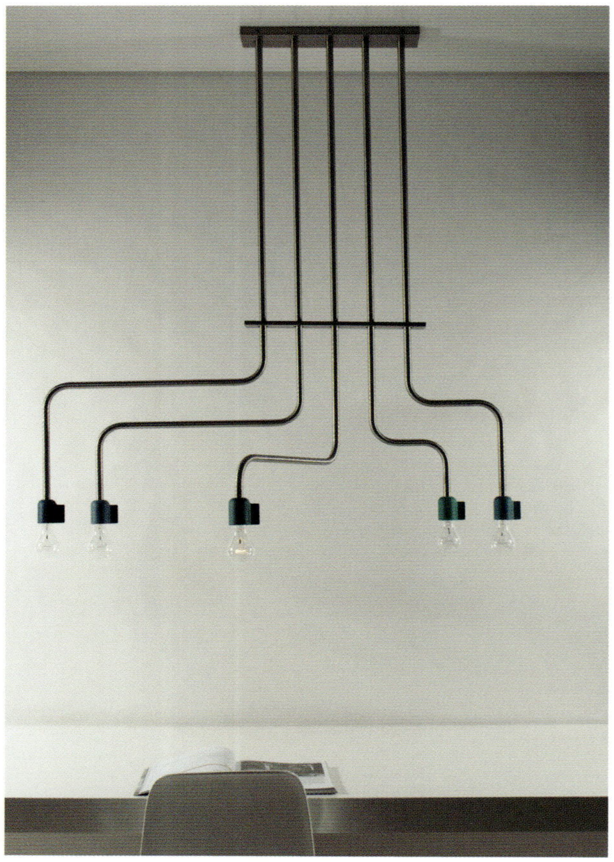

James Irvine: *"I guess the arms rotate. It is a new idea that is reminiscent of a classic chanderlier."*

"When I first read the news about the award I was riding the metro. It was very surreal at that time, but I had the biggest smile ever on my face. Winning this award means a lot to me. It is a recognition of my work. It's also great that this came together with two of my designs going in to production."

industrial design + product design

Design
Yi-Chun Chou, Kuo-Shu Chen
Tainan, Taiwan

National Cheng Kung University
Tainan, Taiwan

RESCUE SEAT
LIFEBUOY

The "Rescue Seat" is a new type of water-rescue equipment that provides additional support to victims waiting for the rescue team. It supports their weight and gives them a more secure feeling. A rubber band around the outside of the life preserver gives the user a place to hold on tight. It can also be used as a handy way to transport and store the life preserver and to get it into position quickly in an emergency situation.

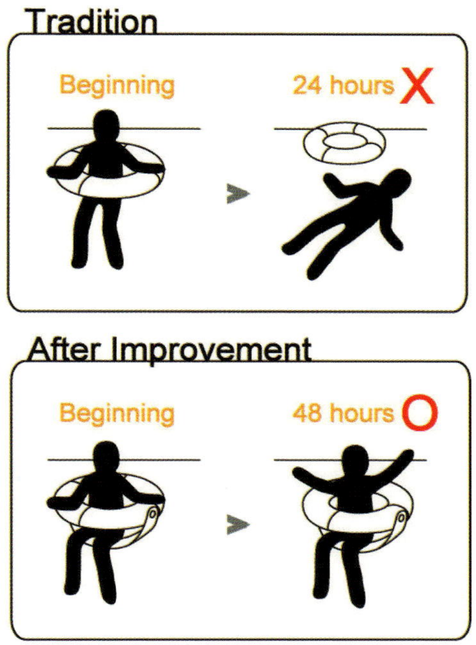

Jeremy Sun: *"A simple yet effective solution that could significantly improve the life-saving function of lifebuoys."*

"It is really exciting to get the iF concept design award 2012. It is a wonderful experience and we are both really proud that we made it. The iF award is the most famous award all over the world. This award made us believe we can conquer any obstacle and has given us more confidence to face new challenges in the future."

industrial design + product design

Design
Jiheon Kim
Seoul, South Korea

Seoul National University
Seoul, South Korea

ROLY-POLY TOOTHBRUSH
TOOTHBRUSH

This toothbrush stands up by itself, just like a roly-poly so that you can put it on a charging plate instead of putting it in a cup. This makes it more hygienic and it can be put down virtually anywhere on a level surface. A built-in ultraviolet lamp is used to sterilize the bristles.

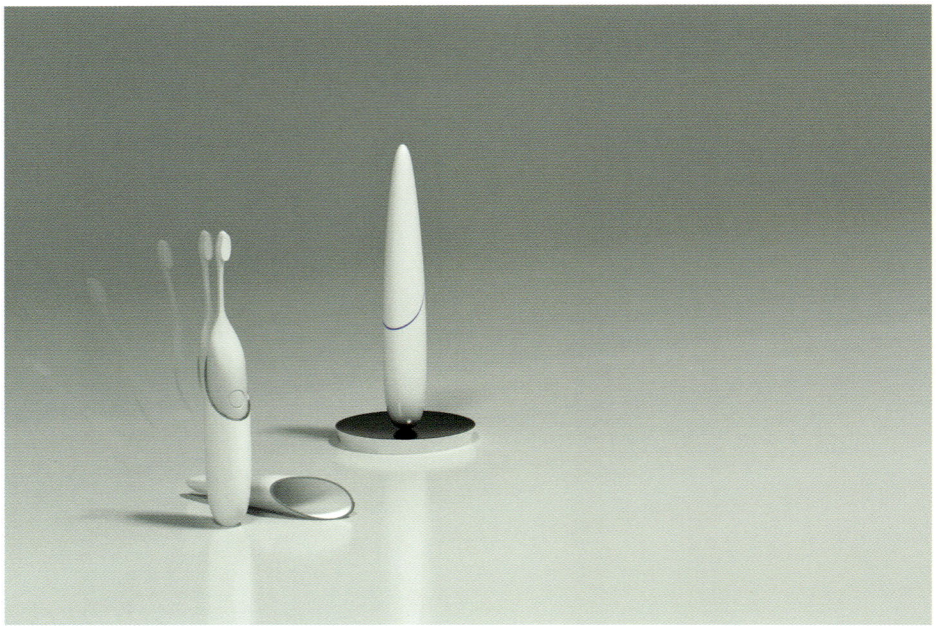

Peter Kuchnicki: *"What fun! A toothbrush that wobbles! And with its cap secured, it's a delightful sculpture. Superlative industrial design."*

"First, I'm very thankful to receive this award. This has been immensely rewarding. And I shall see you some time again as a product designer. Not just a student. Thanks again, and I wish you all the best."

industrial design + product design

Design
Moonjeong Choi, Haeryung Lee, Sujin Lee
Seoul, South Korea

Sookmyung Women's University
Seoul, South Korea

SAVE TRIGGER
LIFEGUARD PRODUCT

"Save trigger" is a gun-like, life-saving device that is used to shoot out a life line to a person in an emergency situation, such as after a water or swimming accident. Objects like life preservers can also be attached. This means that someone's life can be saved with a single shot.

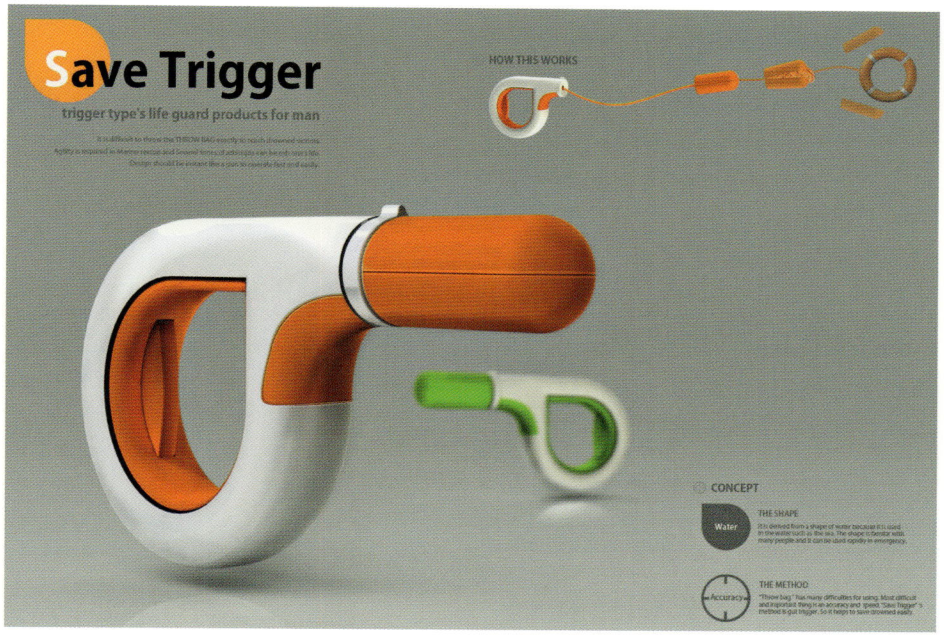

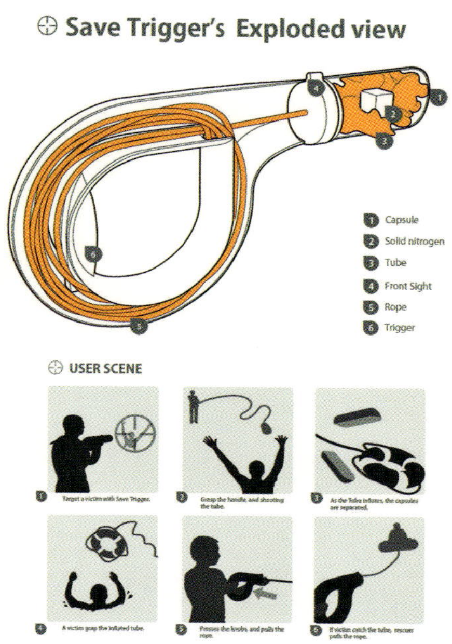

Rolf Schweizer: *"Great idea with a very sophisticated product design and presentation."*

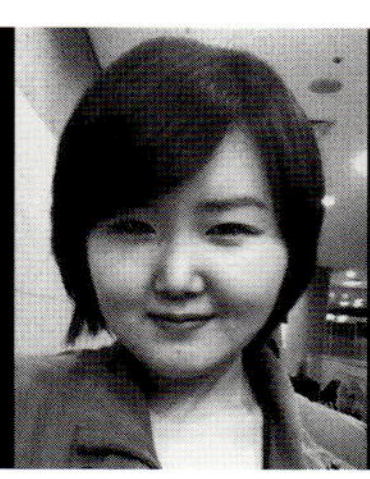

"First, we are very honored to win the iF concept design award and very happy about the results of all our hard work. We will keep trying our best and this award will give us a good opportunity to pull the TRIGGER on our dreams."

industrial design + product design

Design

Kyung-ho Jang, Sung-mi Kim
Daejeon, South Korea

Hannam University
Daejeon, South Korea

SIMCHEONG
PRINTER & SCANNER

"SimCheong" is a traditional fairytale figure in Korea, a devoted daughter who opened her father's blind eyes. This story was the motivation for developing this product: a printer combined with a scanner that is able to read Braille so that blind people can have more social contact to the rest of the world.

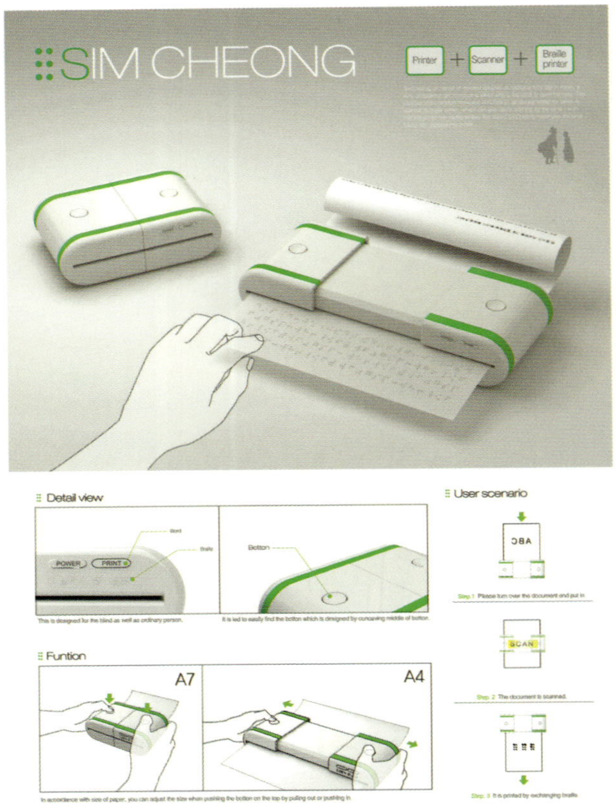

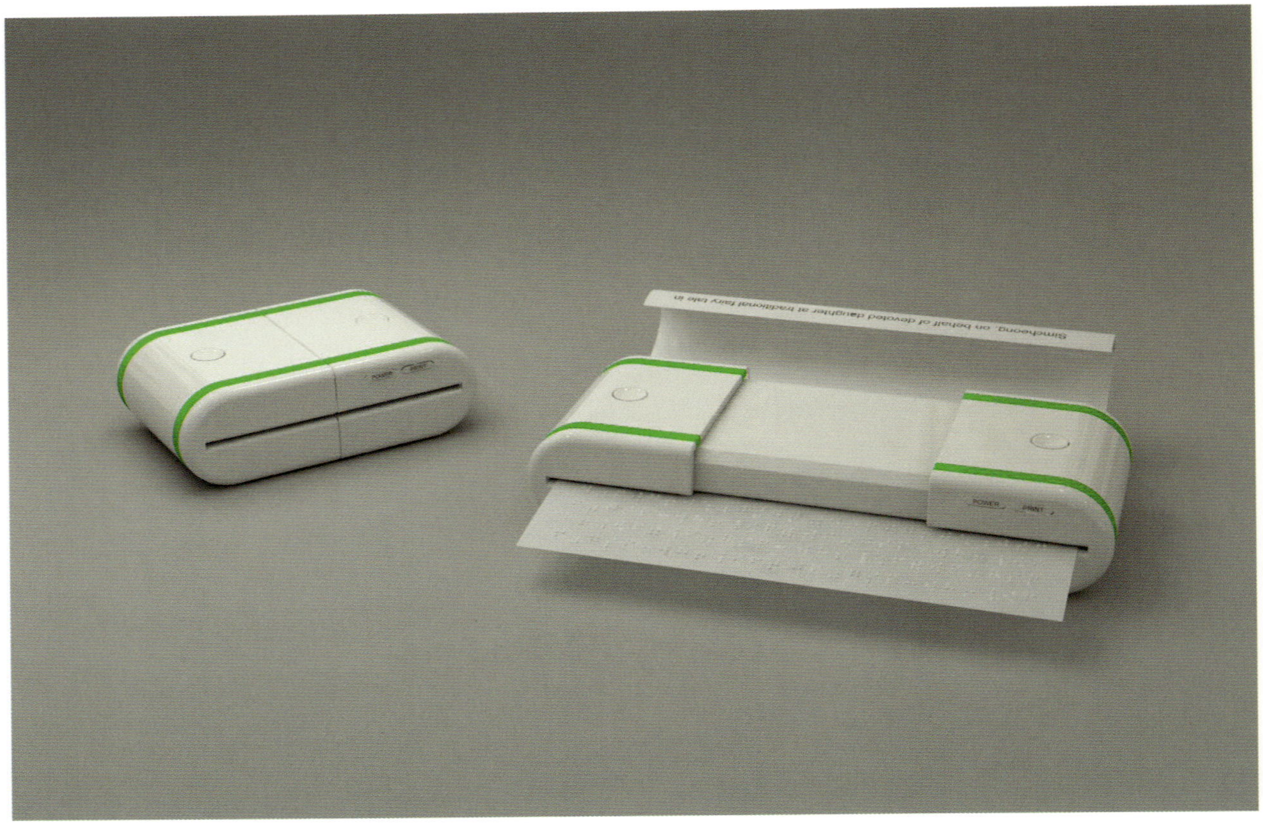

Lidan Liu: *"Froim an emotional point of view, a useful and viable idea to help the blind."*

industrial design + product design

Design

Chih Wei Lai, Chi Wang, Cheng Wei Wang
(Ming Chi University of Technology),
Jia Yu Hong (National Taiwan University of Arts)
Taipei, Taiwan

Taipei University of Technology
Taipei, Taiwan

SMARTEXIT
EMERGENCY EXIT

In the event of fire, it is very important to know how to get to the exit door quickly. However, it is not always safe to open that door because you don't know whether or not there is smoke and/or fire on the other side. This exit door includes a smoke detector with a guiding light. The light changes color to warn people not to enter when the smoke on the other side is too thick and stays brightly lit for 180 seconds to indicate the right direction to go to find an alternate exit.

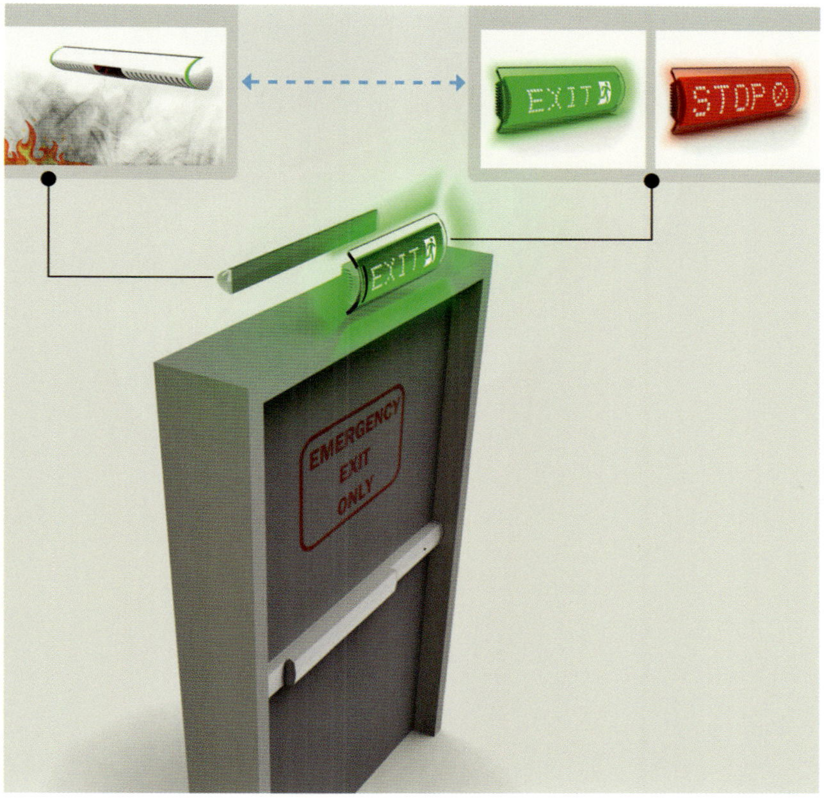

William Wu:
"This design can possibly save a lot of people who are stuck in a burning building. Designers could develop more innovations to protect life."

"First, 'attitude determines everything'; second, 'the opportunity is given to the person who prepares'."

industrial design + product design

Design
Shang-Che Wu
Hsinchu, Taiwan

National Chiao Tung University
Hsinchu, Taiwan

SOUP-BAG
PLASTIC BAG

The "soup-bag" is a plastic bag designed for take-out bowls of soup. Unlike traditional T-Shirt bags with a scrap of plastic for the handles, the soup-bag uses this area as another bag. As a result, the soup-bag reduces production waste and its trapezoid bottom prevents the bowls from spilling.

"I really appreciate winning this prize."

industrial design + product design

Design
Wang Can
Beijing, China

Beijing Institute of Technology
Beijing, China

SOURCE OF WATER
WATER PURIFIER

After a flood, one of the most terrible natural disasters, it is often very difficult to find clean drinking water. "SOURCE OF WATER" is a water filter specially designed for refugees in flood-stricken areas to help them get clean drinking water quickly and conveniently. It is easy to use and based on a simple principle: water pressure, active water purification material and an ultra-filtration membrane are used to purify and decontaminate water automatically to produce clean drinking water. It is also readily portable and can be put into mass production at a low cost.

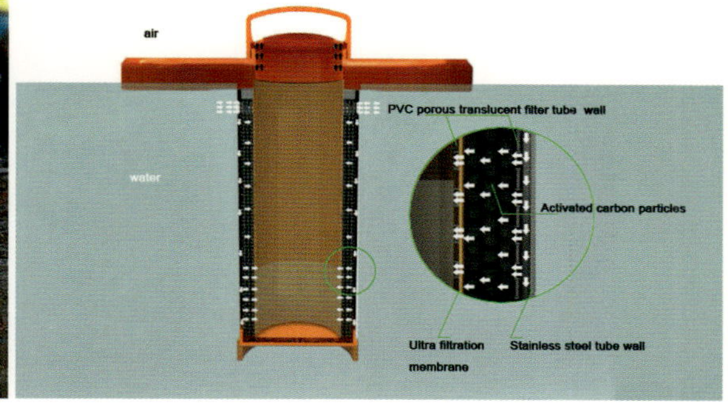

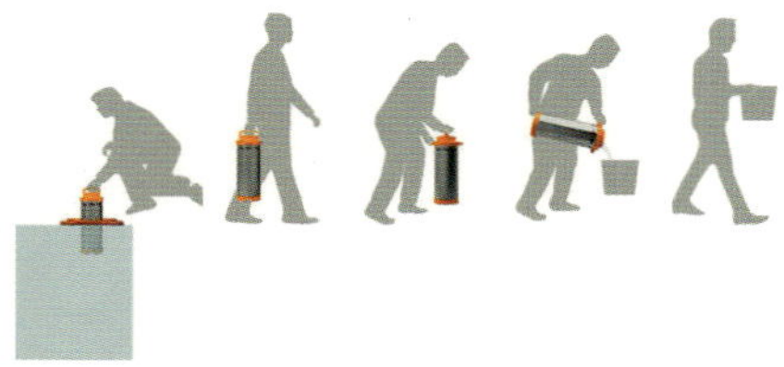

Prof. Claus-Christian Eckhard: *"If it works like the "LifeStraw", it can be an essential contribution to situations where fresh/drinking water can save lives. The need for the cross-shaped float is not 100% clear to me and also how to keep it in place, so that the entire filter is not carried off course."*

"It was a big surprise for me to get this special award. I can't wait to tell my families and good friends. The award gives me great confidence and strength to go on walking along the road of design. I will continue to work harder and try my best to help people who are in need of help through my design! That's my dream!"

industrial design + product design

Design
Roger José Mafra Burda
Curitiba, Brazil

Pontifícia Universidade Católica do Paraná – PUCPR
Curitiba – PR, Brazil

SPHYGMO HAND
MEDICAL PRODUCT

This product was developed for use by the elderly and health professionals. The elderly often have difficulty in putting on the cuff; this model has been equipped with a bracelet inside the cuff that automatically wraps around the arm for easier handling. Another big advantage is that the medical staff no longer needs to carry around the actual device. Instead, they can wear it like an armband with a badge to identify themselves to be able to work more quickly and easily and to help save more lives.

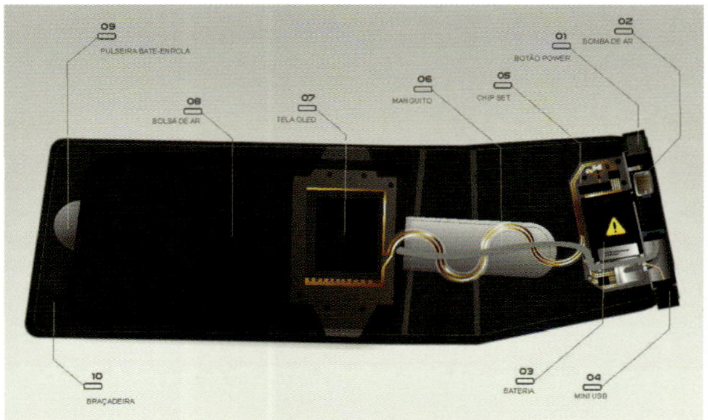

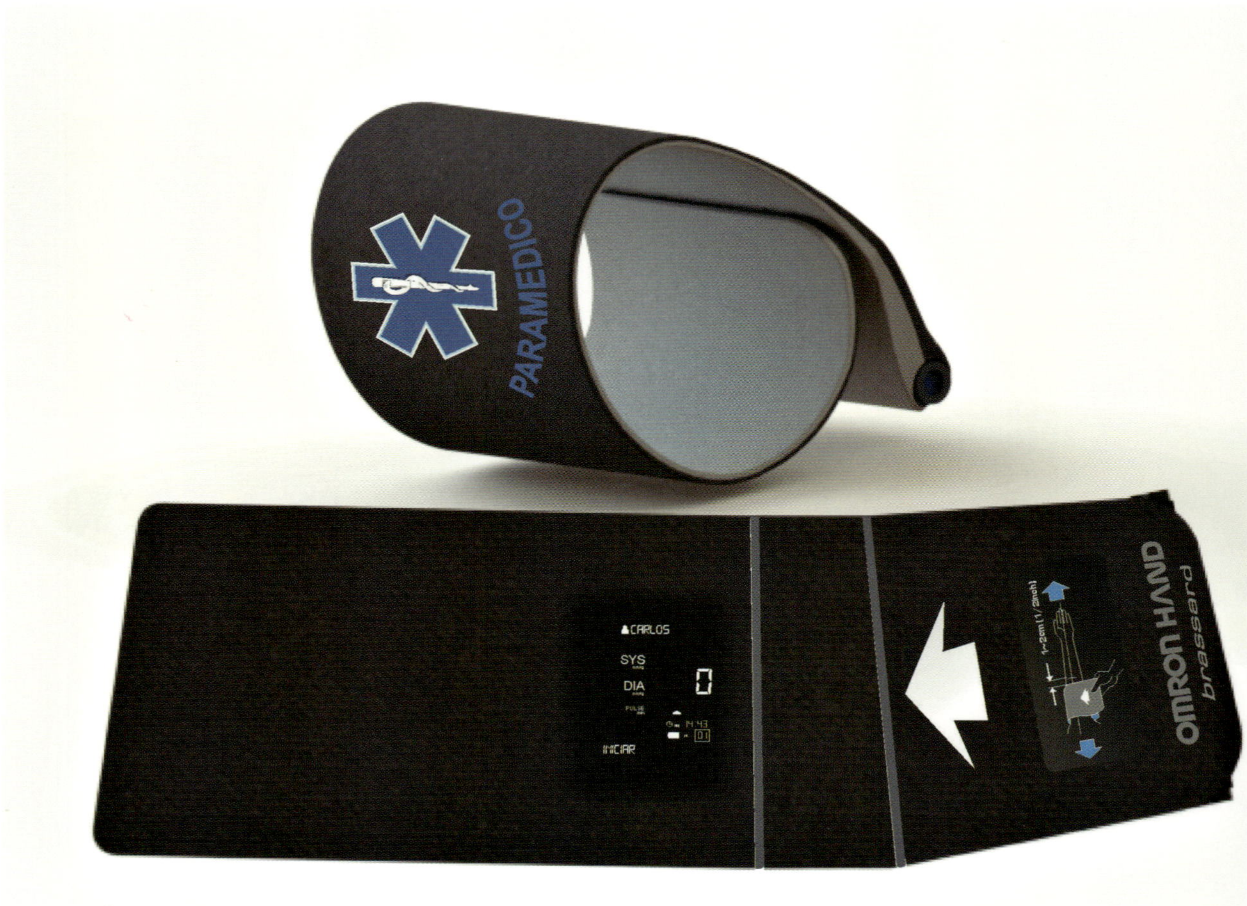

Zlatko Kapetanovic: *"Design and aesthetics support functionality and simplified usage for the elderly population and it has potential to prove itself as a life saver."*

"The iF concept design award is proof of the power to transform the world with ideas. We need incentives to believe in our own dreams. This award represents the recognition and realization of our dreams."

industrial design + product design

Design
Yen Chang Chiu, Chun Wen Wang
New Taipei City, Taiwan

Shin Chien University
Taipei, Taiwan

ST- BRACELET
ELECTRONIC PAPER

When you go to the hospital, you normally have to register first and then wait to be examined. While the registration receipt estimates the appointment time, it is often delayed or moved up depending on various situations.
ST-bracelet is made of flexible e- Paper (with flexible sheet metal) and displays the number of the current patient. The user can then estimate the time remaining until his/her appointment and use the waiting time efficiently. After the examination, the patient can pick up the prescribed medication with the same bracelet, the bracelet is a sustainable solution that can be used again and again and saves paper.

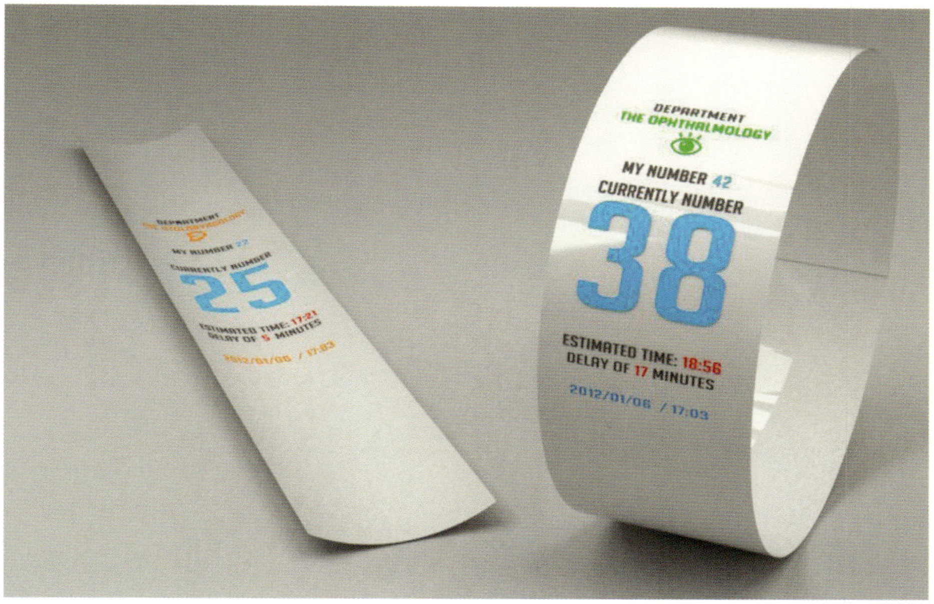

J'aime Cohen: *"This is truly inventive and well designed."*

"We will keep our passion of design forever."

industrial design + product design

Design

Jia-Weih Ding
Taipei, Taiwan

National United University
Miaoli, Taiwan

STABLE HAND TOILET

Studies say that 90% of home accidents occur in the bathroom. This innovative toilet has been designed to meet the needs of the elderly and arthritis patients in particular. The sturdy hand rails made of stainless steel provide the necessary support when using the facilities.

"After a traffic accident I was involved in two years ago, everyday activities, even going to the toilet, became much more difficult for me. I developed this concept to give myself greater independence and now it's been approved! The iF concept design award has encouraged me to continue to design for people with disabilities. I really feel very honored and sincerely appreciate everyone who contributed to this success."

Design
Florian Schmid
Penzing, Germany

Hochschule für angewandte Wissenschaften –
Fachhochschule München
München, Germany

STITCHING CONCRETE STOOL

"Stitching Concrete" is a project, inspired by the contrasts in materials of concrete and canvas. These stools are made by folding fabric that has been impregnated with cement, and then soaked in water. A wooden mold supports it while it dries and hardens within 24 hours. The material combines the warm softness of cloth with the stability of concrete, while the finished surface retains a soft appearance. It plays with the visual illusion that cloth is not strong enough to sit on. The stools are designed for use indoors and out. The material is UV and chemical-resistant, fireproof and water-resistant.

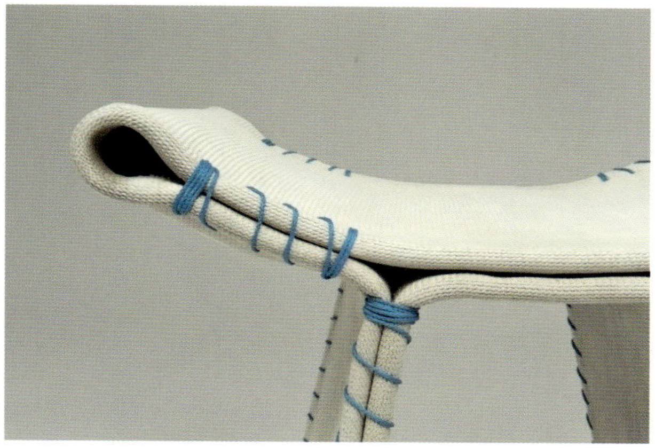

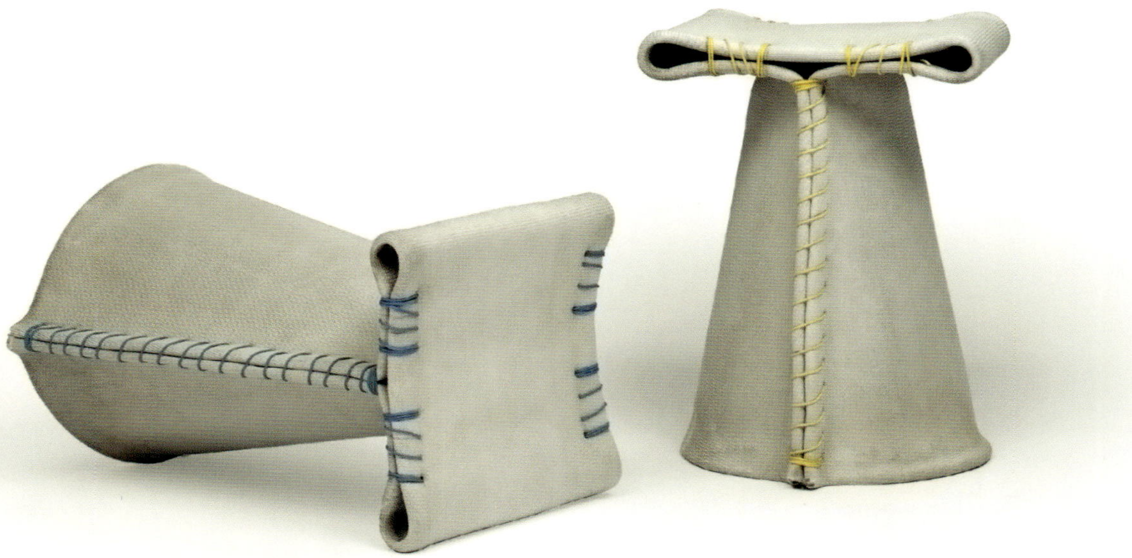

Boudewjin Soetens: *"A beautiful example of innovative material use with unique results."*

"First I started to run through the house and shouted it out loud.
The iF concept design award was something I always wanted to achieve in my dreams and now it came truth. It is a great honour to have that and it will surely encourage me to go on with my design career."

industrial design + product design

Design
Jae Sun Yoon, Jung su Kim
Seoul, South Korea

Seoul National University of Science and Technology
Seoul, South Korea

TSUNAMI SHELTER
EMERGENCY SHELTER

The "Tsunami Shelter" is a product designed to save the lives of people surprised by a tsunami, a sudden flood or a tidal wave from the sea. It can be set up very quickly and is equipped with a floatation device, a buoy and a device that emits an SOS message and a light to make the shelter more highly visible to a rescue team.

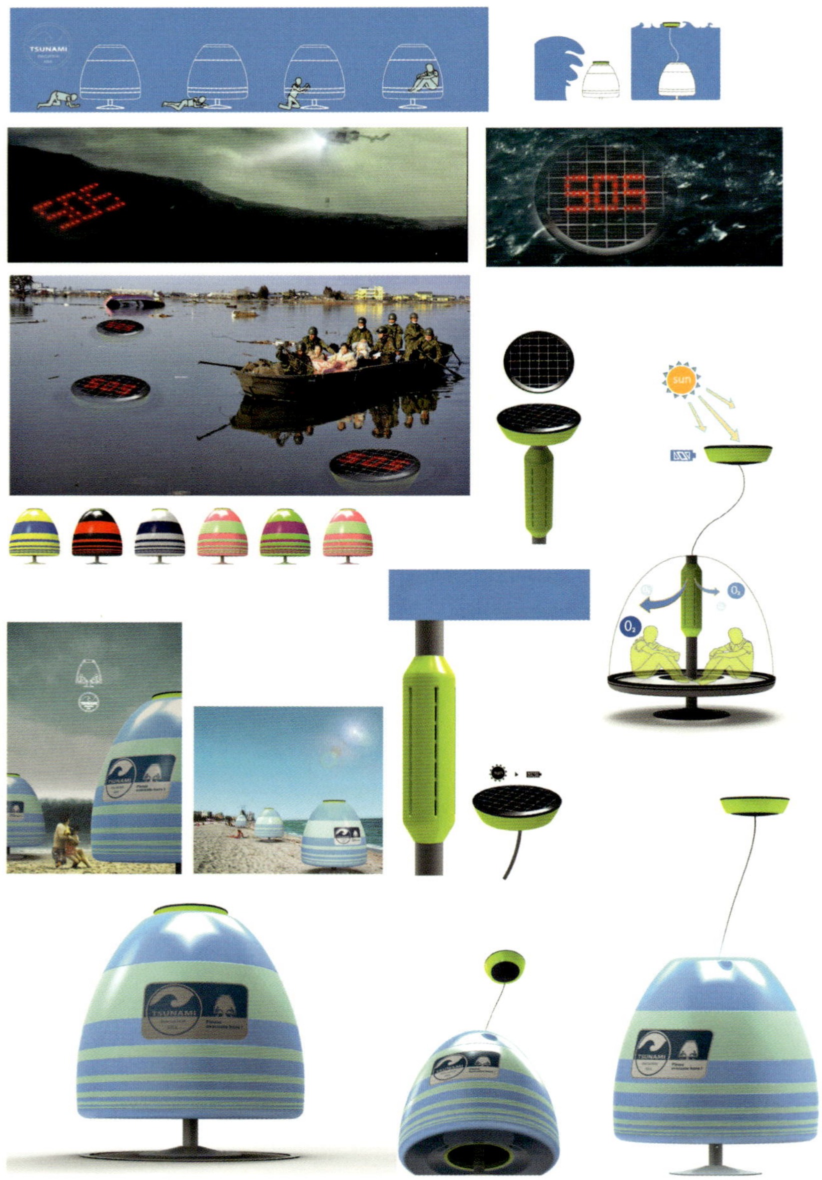

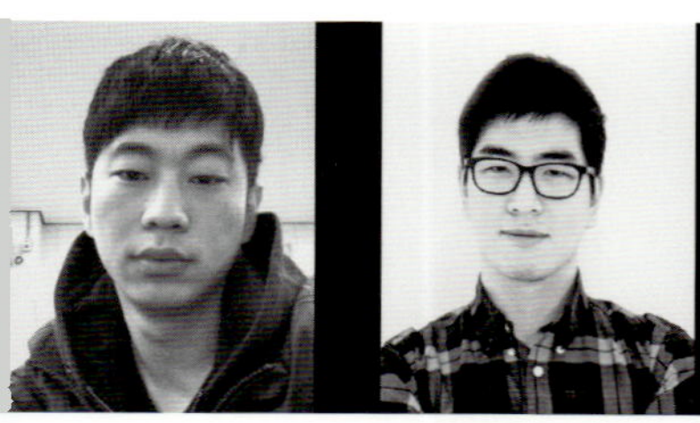

"We believe that the fundamental approach in regards to design is to provide the service of giving a sense of personal warmth and enjoyment to people who are surrounded by dry and dull spaces through incorporating changes in design. We want to create designs like the works of Victor Papanek which are rational and are humanistic and are for people who are marginalized in society."

industrial design + product design

Design
Li Yin-kai, Cai Fu-yu, Syu Shou-ren, Fan Cheng-kuei
Taipei, Taiwan

National Taipei University of Technology
Taipei, Taiwan

WAVE PAINTING BUCKET
PAINT BUCKET

Because a wave shape has been used for the paint can and the lid, you can open the lid without having to use any tools. Thanks to the wave shape, the lid can also be easily picked up if it flips over without getting any paint on your hands.

Lidan Liu: *"Smart solution to the problems of a conventional paint bucket. Reasonable, easy to use, ergonomically correct, and also includes a high aesthetics aspect."*

"It's a great honor to receive an iF concept design award 2012. From the original observation to the finished model, it took a year to complete this product design. It is not easy to stand up against more than 10,000 other entries. We look forward to our design being mass-produced to provide a more convenient bucket."

industrial design + product design

Design
Tenggao Shi,
Yujuan Zou (Zhejiang University),
Danshuang Li (Zhejiang University),
Limin He (Zhejiang University)
Hangzhou, China

Zhejiang Gongshang University
Hangzhou, China

WING
TRANSPORT ROLLS

This is a design for dual use. First of all, it can be used as training wheels on a kid's bike and helps an adult to support the child who is learning to ride. Secondly, it can serve as a shopping trolley, which is particularly useful for elderly people who cannot or do not want to carry heavy bags.

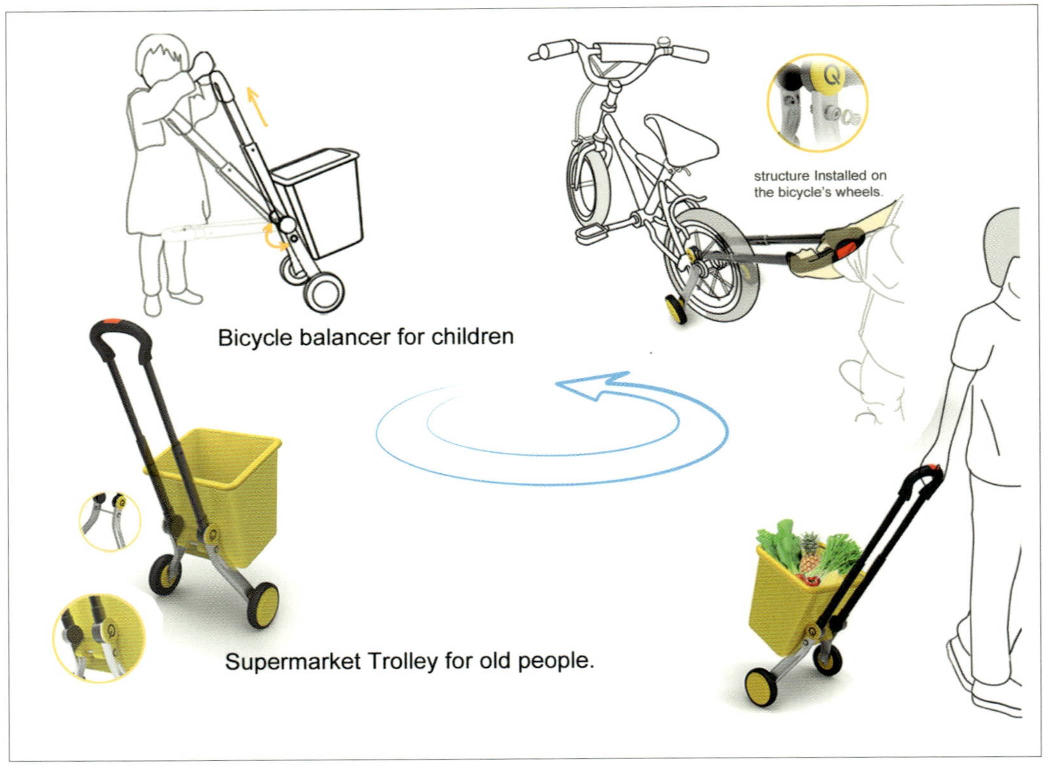

"We are honored to take part in the iF concept design award 2012 as students. Our team and our universities are truly delighted to win this prize, really honored to stand out from 10,566 other qualifying entries. Thanks to the iF organizing committee for giving us such an opportunity. We hope to make the world a better place through design."

communication design + multimedia design

Design
Jiu Seo
Seoul, South Korea

Inha University
Incheon, South Korea

DAILY RECEIPT
ART JOURNAL

This is a journal in which daily receipts are collected to tell the story of what happened that day, making it possible to analyze the actions, thoughts and other things on any particular day. With our fast-paced lives today, this is an easy way to keep track of what we do and where we have been as a reflection of our own lives.

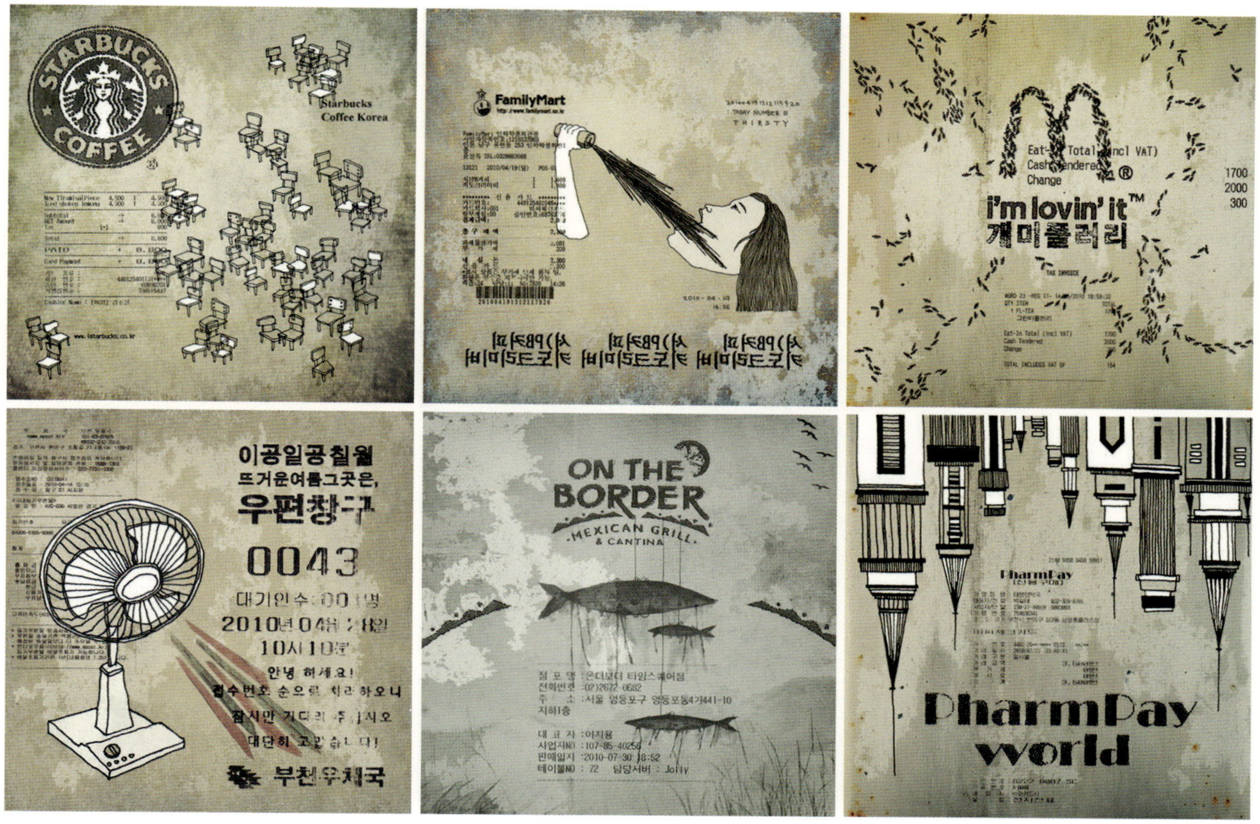

Martin Köttering: *"This type of journal using "Daily Receipts" archives shopping receipts to focus on a common problem that affects many people: I shop, therefore I am! It's a touch-in-cheek, yet critical look at how our shopping affects our identities. A successful design approach that encourages us to reflect on what we spend and where every day."*

"I'm really honored to be an award winner and I'm glad to have the chance to show my work to an international audience. In the future, I want to concentrate on more projects that focus on increasing and improving communication."

communication design + multimedia design

Design
Justine Kaczmarek
Hannover, Germany

Fachhochschule Hannover
Hannover, Germany

GABE?DEFEKT?
FIVE MAGAZINES

My work deals with the Savant Syndrome, a rare but spectacular condition, in which people with various developmental disabilities have astonishing islands of ability. These skills stand in stark contrast to their handicap and involve music, art, mathematics and language – in all cases, they are linked with phenomenal memory. Savants have a different perception of their environment. Their thinking processes and inner pictorial world are dominated by patterns rather than images. They are "pattern thinkers". Five magazines, introducing a Savant and his/her ability, give a visual interpretation of what this individual perception may look like.

Sybs Bauer: *"It is a interesting idea to look through the eyes of what we consider as a handicapped person. Without a doubt, this could shift perspectives in various directions."*

"I entered my project in the iF concept design award to see whether or not it would stand the test outside of the college environment. Being one of the best 100 is an unbelievable motivation boost for me and for the work on my Master's thesis."

communication design + multimedia design

Design
Wei-Je Huang
Tainan, Taiwan

Tainan University Of Technology
Tainan, Taiwan

Jan Stavik: *"A charming and easily understood warning message to us all."*

LOCH NESS TRUTH
POSTER DESIGN

This work is a poster on the topic of global warming. The monster of Loch Ness looks different as it rears its head from the Loch because of the rising water levels due to global warming. The work uses an elephant's trunk and the monster's body as we imagine it to demonstrate a unique visual experience and to make a lasting impression. The seemingly Western color scheme used in this work is actually the "leaving blank" effect of Oriental landscape painting. The structural surface of the poster reflects the width and depth of the Loch. The floating elephant in the visual center symbolizes the issue of global warming.

"I am very happy to receive this award and my thanks go out to my teachers and family. I hope that this work will convey my concern about the crisis of global warming and encourage people to re-think their attitudes towards their impact on the environment and other people."

communication design + multimedia design

Design
Yi-Shan Hsieh
Taichung, Taiwan

Shih Chien University
Taipei, Taiwan

TEA CEREMONY
TEA BAG

This concept uses tea written onto paper, describing the tea ceremony and sharing the beautiful harmony of this tradition with the world. You simply tear off the tea page, read its message and then add boiling water to brew a delicious beverage.

Elaine Ann: "A very poetic presentation of tea, creating a new cultural experience."

"I appreciate this recognition from the iF concept design award. This honor is really encouraging me to continue observing more details in my daily life. To try and find the balance between traditional approaches and new techniques and to use design to create a better world. Design makes the world go round."

communication design + multimedia design

Design
Lydia Scharlata
Neuss, Germany

Hochschule Niederrhein
Krefeld, Germany

Peter Kuchnicki: *"A beautiful example of information design, and what an intriguing subject. The layouts, colors and charts all combine to clearly communicate. Lovely!"*

ZWEIUNDZWANZIG°
BOOK – SYSTEMATICALLY GETTING LOST

Getting lost is one of the primal fears of the human race. The fear of losing control makes us blind to the possibilities that could result from being disoriented. This book is a travel guide that teaches readers how to get lost by choice – but always with the option of finding their way back without using devices like maps or navigation systems. Divided into a theoretical and a practical part, the book teaches readers how to interpret the urban environment like a boy scout would interpret the forest. Way-finding strategies are combined with on-site practice to explore the diversity of a city in a new way.

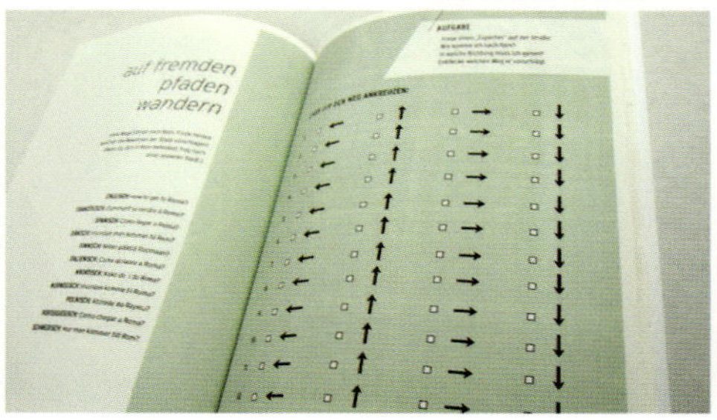

"When I first heard that I had won, I could hardly believe it. Winning this award is a real honor. But not only that: it is the crowning finish to my studies and gives me even greater motivation to launch my career in the field of design and continue working on the quality of my projects."

Design
Yakun Zhang, Yu Liu, Danwei Ye
Rochester, United States of America

Rochester Institute of Technology
Rochester, United States of America

BACKBONE
BACKPACK

Sometimes it's difficult to find something you need when you are carrying a large backpack. "Backbone" is designed to fit underneath your backpack against your back to allow you to attach small separate containers at different points. It is stylish, made of high performance plastic and is multipurpose. You can use the clip hooks to attach plastic bags as your containers or use rope to tie on firewood when you go hiking.

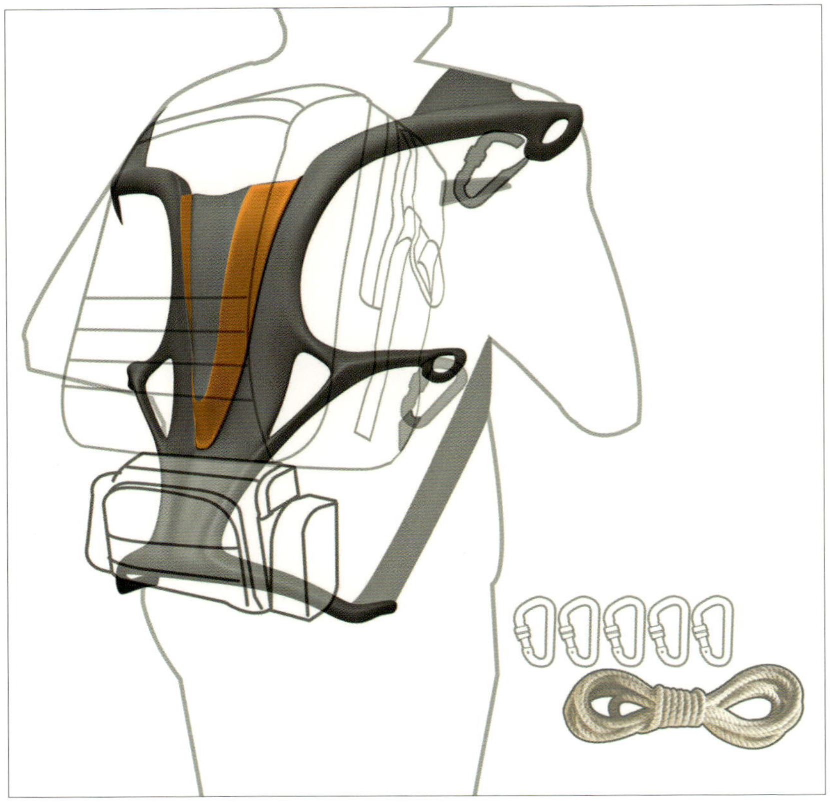

Angela Carvalho: *"A solution based on ergonomics and the concept of flexibility. Makes it possible to transport different types of loads and bags. We can say that the product is reduced to its essentials."*

"We are very glad that we won the iF concept design award and thank the whole iF community for their appreciation of our work. We are so happy, not only because our hard work has been rewarded, but also because it represents a milestone in our careers. We can hardly believe that both our entries won. As industrial designers, we will continue to do our best to develop more great ideas. Thanks for all of your support."

interior design + architecture

Design
Eun Whan Cho, Christiana Papadaki, Sin Yub Hwang
Seoul, South Korea

Domus Academy
Milan, Italy

BLU
EXHIBITION

"Blu" is an experimental project using light that is meant to evoke emotions and sensations to create two types of feelings, namely discomfort and hope. The tension between these two feelings is illustrated by the water element, which is not used in its physical form, but more for its psychological effects. When water is seen at different levels and not in the way sea water is usually seen, it becomes possible for visitors to 'develop' a story of emotions and to experience these emotions in the four stages of water and light as they move through the gallery.

"Because we worked with good partners, it was possible for us to achieve good results and to enjoy working together.
In addition to that, We would like to express our sincere gratitude to Antonella Dedini and Cinzia Ferrara who were our academic supervisors for this project design."

interior design + architecture

Design
Wen-Jun Cheng, Mei-Ling Lin, Qian-Yu Dong
Tainan, Taiwan

Tainan University Of Technology
Tainan, Taiwan

ORGANIC FABRICATION
ORGANIC FARM ARCHITECTURE

This building, located in hilly Kuan Miao Hsiang, has been restructured using the existing architecture and including bamboo weaving to remodel its spatial atmosphere. The covering with earth reflects a green architectural building approach. This, plus topographical features and the collection and reuse of rooftop rainwater, makes it an organic artistic farm integrating energy conservation, environmental protection, and cultural heritage.

Jan Stavik: *"A fascinating, futuristic environmental concept with a clearly local orientation."*

"We were so happy and surprised to learn that we had received the iF concept design award 2012; it is truly a great honor for us to be among the best 100. We give our heartfelt thanks to everyone for their tremendous support and assistance. We are highly motivated and inspired and want to invest our passion in providing design to our community. Thanks iF."

interior design + architecture

Design
Hyunjung Kim
Daejeon, South Korea

Korea Advanced Institute of Science and Technology – KAIST
Daejeon, South Korea

SHADE PIXEL
PHYSICAL DISPLAY

"Shade Pixel" is a physical display that uses a shapeable skin to create concave pixels to represent such information as patterns, numbers and texts. Unlike typical light-emitting displays, "Shade Pixel" uses shade to visualize information in a manner similar to Cuneiform or Sunkenrelief. This evokes a primitive naturalness and provides sensory and hence emotional comfort to users. In addition, its non-luminescent nature and simple appearance reduces unnecessary visual noise, especially when it is not in use.

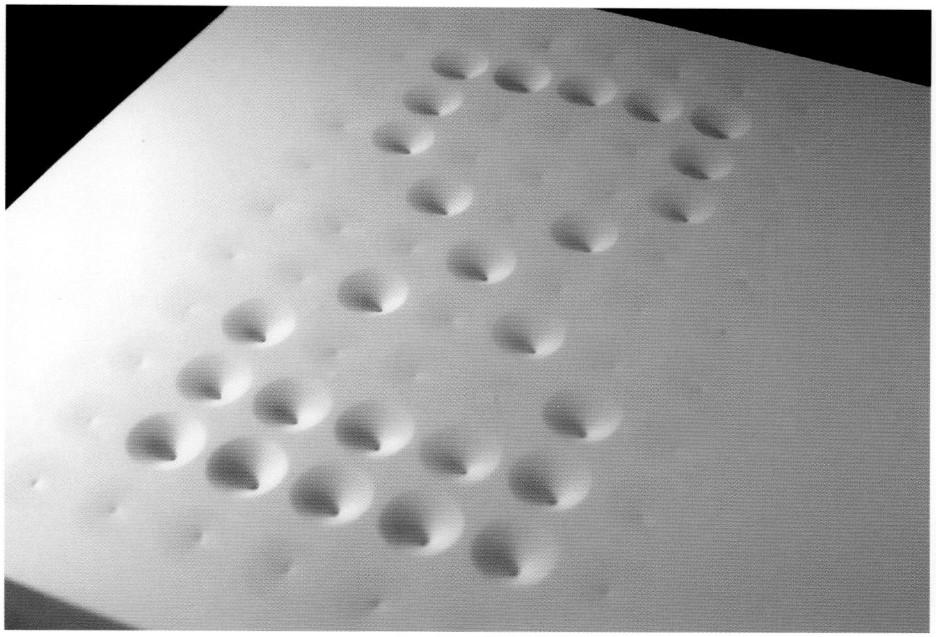

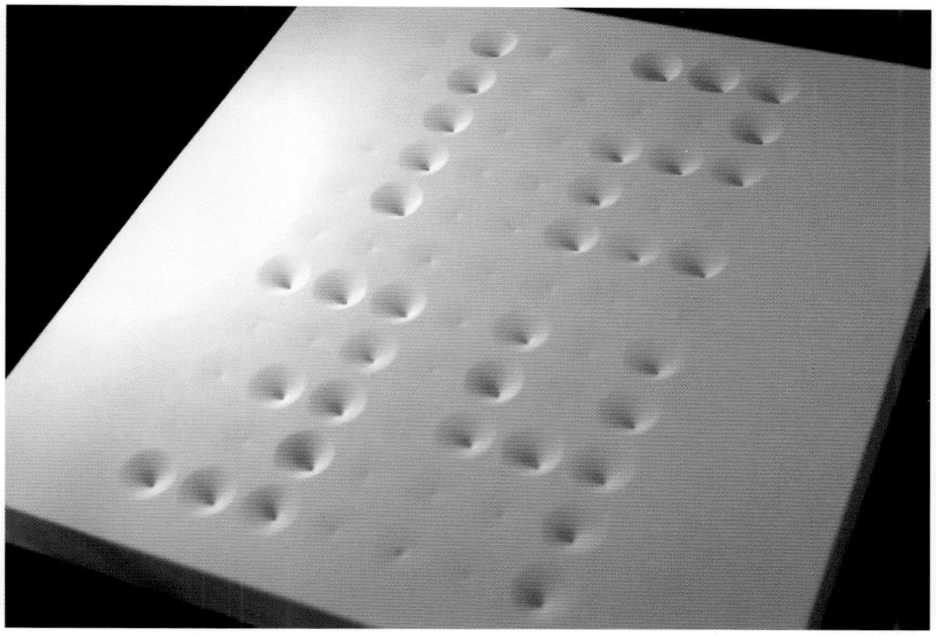

"There are no words to describe how happy I am right now. It is my great honor to receive this award. It definitely gives me the encouragement to continue pursuing my dreams. My sincerest thanks to everyone involved. This is just the beginning."

SPECIAL PRIZE: HANSGROHE PREIS 2012

For the second year the iF concept design award included a special prize. This year it was entitled "Hansgrohe Preis 2012: Efficient Water Design - My Green Shower Pleasure". This prize focused on the topic of the sustainable use of water in the bathroom environment.

Of around 150 entries competing for this special award, six won. Three won the special prize and shared a purse of EUR 3,500 in prize monies. Three further concepts received a special mention and EUR 500 each in prize monies.

On the next pages we present the jury and the award winners of this special prize.

"The responsible management of water, one of life's most basic elements, has been a major concern with us at Hansgrohe for a long time. With the Hansgrohe prize 'Efficient Water Design', we want to encourage young designers to focus more on this topic and to develop more creative ideas. The winning entries show that good design is more and more about aesthetics that can also have a positive impact in terms of saving resources and our environment."
PHILIPPE GROHE, AXOR/HANSGROHE SE, SCHILTACH/DEUTSCHLAND

hansgrohe

2012

JURY SPECIAL PRIZE

ANDREAS HAUG
JAN HEISTERHAGEN
DR. CARSTEN TESSMER

Andreas Haug (*1946) is – together with Tom Schönherr – a managing partner of Phoenix Design Stuttgart, a well-known design studio in Stuttgart. After completing his studies as a Product and Industrial Designer at the State Academy of Art and Design in Stuttgart, he became the first employee at Esslinger Design in 1972. Until Phoenix was founded in 1987, he worked as a Design Consultant and Vice President at frogdesign.

ANDREAS HAUG

JAN HEISTERHAGEN

Jan Heisterhagen (* 1969) has been the head of Product Management at Hansgrohe AG in Schiltach, Germany, since 2008. Before that, from 1996 to 2000, he worked for Vitra GmbH in Weil am Rhein – first as a Junior – later as a Senior Product Manager Product Manager at Vitra International AG in Basel, Switzerland. He made the move to Hansgrohe in 2001. He studied industrial engineering at the Offenburg University of Applied Sciences and Arts. The certified project management expert has two children, 11 and 2 years old.

DR. CARSTEN TESSMER

Dr. Carsten Tessmer (*1963) is the Corporate Spokesperson and head of Corporate Communications at Hansgrohe AG in Schiltach, Germany. After completing his studies at the Università degli Studi di Milano (Italy), he ended his studies of Political Science with a Master's degree at the University of Mannheim in 1992. Before he began to work for Hansgrohe in 2004, he was, among other things, employed at the Federal Chancellor Willy Brandt Foundation in Berlin and at Zumtobel Staff GmbH in Dornbirn, Austria.

industrial design + product design

Design
Ahmet Burak Aktas, Salih Berk Ilhan,
Burak Soylemez, Adem Onalan
Ankara, Turkey

Middle East Technical University
Ankara, Turkey

Prize money EUR 1,500

IWASHIT
SHOWER-WASHING MACHINE

"iWashit" is a product that combines a washing machine with a shower stall to enable users to wash their own clothes with the same water that they used while showering. While the user is having a shower,"iWashit" collects the grey water from the shower stall. After that, all the water in storage is cleaned and ready to use again either by the shower or the washing machine. "iWashit" refills itself from the water mains in case of water loss. This concept brings lots of advantages, both in households and for institutional usage such as fitness centers, festivals or airports.

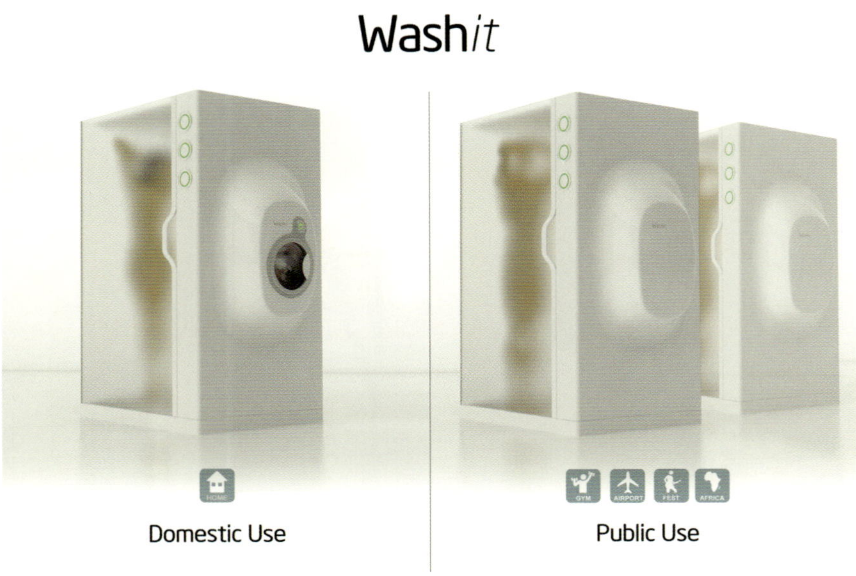

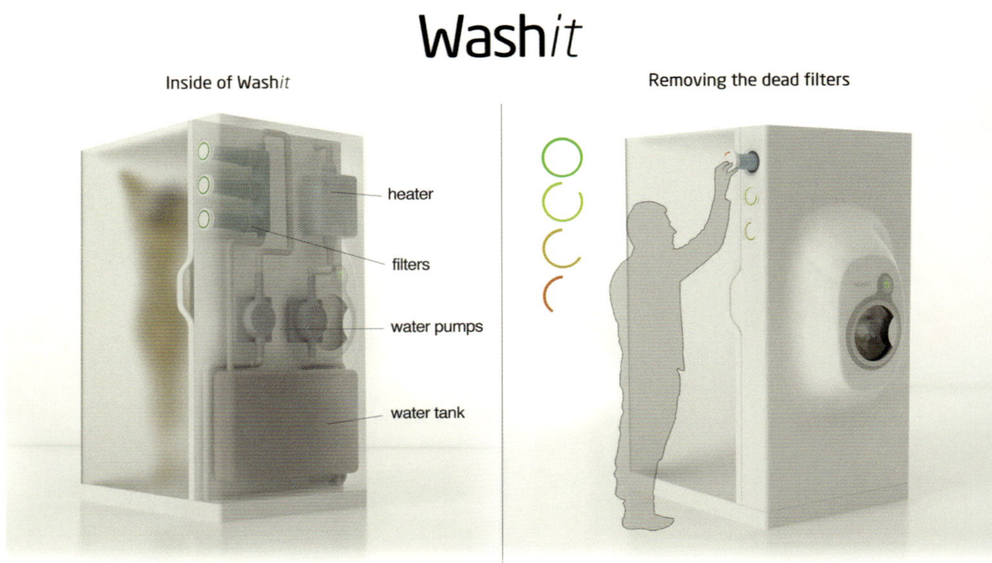

Washit has a closed water system with filters, that uses same water for both showering and washing clothes.

"This is an interesting combination that is very realistic for today and makes sense. A clever solution: identical components, resource-saving, aesthetic appeal, hygienic. Great job!"

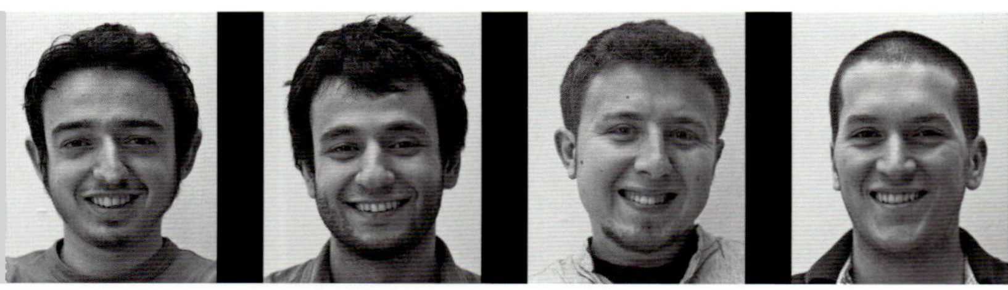

"As young designers, we think that winning this award is a huge thing. It showed us that we are not on the wrong path and all of our efforts are worth making."

communication design + multimedia design

Design
Matthias Laschke
Essen, Germany

Folkwang University of the Arts
Essen, Germany

Prize money EUR 1,000

SHOWER CALENDAR
SHOWER CALENDAR TO SAVE WATER

The "Shower Calendar" to save water is a computer-controlled display mounted on the shower stall. At the touch of a button, a large, colored dot appears on a calendar matrix on the current date. The dot represents 60 liters of water. As the water is consumed, the dot gets smaller until the user turns the water off or the 60 liters have been used up. The dot stays on the calendar as a reminder that the shower was taken.

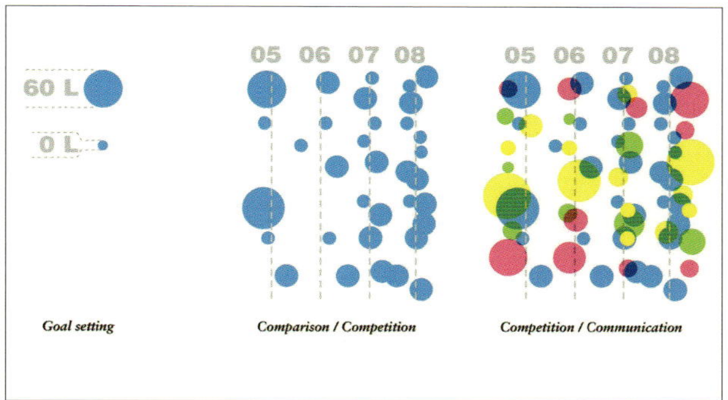

"This is a playful way to heighten people's awareness of their water consumption. The concept reflects high-quality design."

"I am really happy about receiving this award for my work. It shows that design can and should contribute to a better socitey rather than just 'making things pretty'."

industrial design + product design

Design
Jung eun Lim
Cheongju, South Korea

Cheongju University
Cheongju, South Korea

Prize money EUR 1,000

WASH DOME
SHOWER BOOTH

The "Wash Dome" is a "tub" installed at floor level covered with a glass dome that is a closed shower stall, a bathtub, a whirlpool and a sauna all in one. A solution that meets all these needs. The shower wall, which is designed as a kind of "water droplet", stores moisture and provides for a consistent temperature.

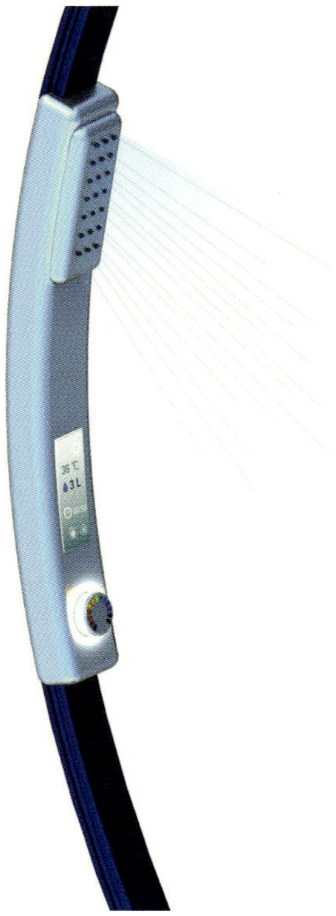

"The multifunctional use is very convincing here: bathing, showering, enjoying a sauna or a whirlpool – everything in one product. The very attractive shower enclosure creates a closed room and a very pleasant new feeling for the space."

industrial design + product design

Design
Sang Un Kim
Ansan, South Korea

Hanyang University
Ansan, South Korea

Prize money EUR 500 (Special mention)

"This is a very simple, but effective solution to control water pressure with the handheld shower head. It has a very aesthetic feel!"

BLUEWHALES
BATHROOM PRODUCT

The "Bluewhales" handheld shower makes it possible for users to individually and easily adjust the water pressure with the palm of their hand. The more pressure that is applied to the rubber-coated shower head, the stronger the water pressure is. Less pressure can be used for face cleansing, for example. This is not only convenient, it also saves a lot of water.

"Thank you God. I can do all things through Christ who strengthen me."

industrial design + product design

Design
Liu Yi, Luo Jing, Jiang Yuning
Hangzhou, China

Zhejiang University
Hangzhou,China

Prize money EUR 500 (Special mention)

AIR CURTAIN SHOWER C
SHOWER CURTAIN

This is a shower stall that uses a shower curtain consisting of an airstream instead of the traditional glass or plastic. It comes out of the floor and prevents any water from spraying out of the shower stall. This "open" shower concept gives users a feeling of freedom. And the circular water jets provide for a new type of showering experience on the one hand and efficient water consumption on the other. When not in use, the bathroom appears much larger.

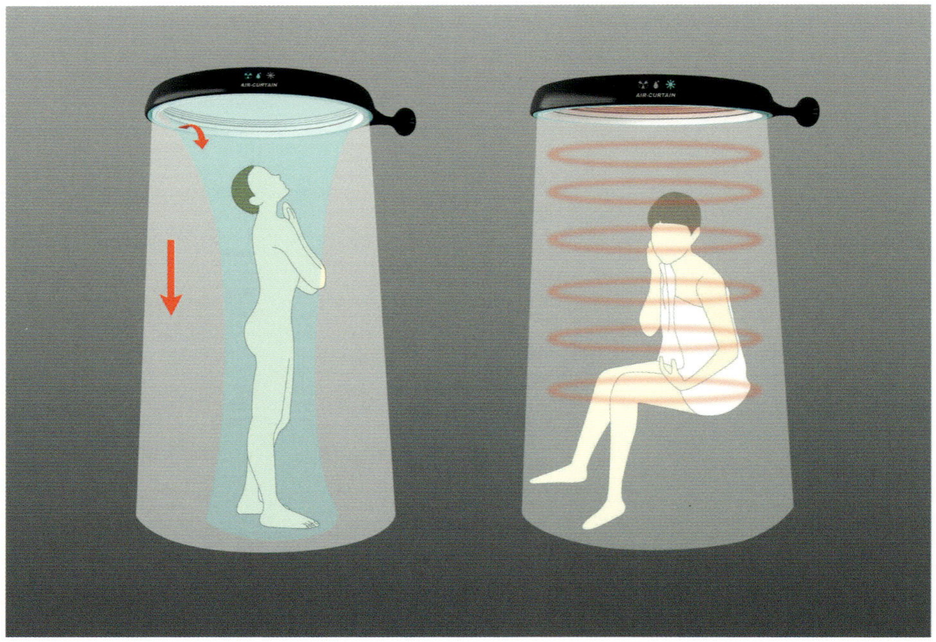

"Delightful how transparent a large shower can be made to appear. Very harmonious connection of the shower head to the wall. The water-saving effect is very interesting."

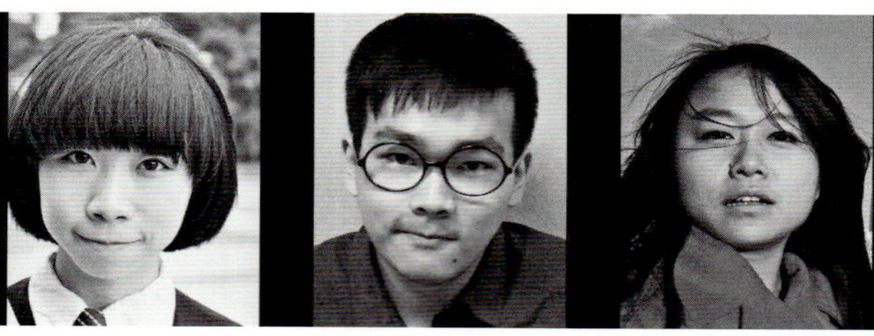

"We felt so surprise when we heard we got an award. It encourages us to continue our hard work and stay focused on design. We believe that innovation design must combine with technology, commerce and user, if we do these, we can make a unique design."

industrial design + product design

Design
Yasaman Sheri
Montreal,Canada

Concordia University
Montreal, Canada

Prize money EUR 500 (Special mention)

LOTA AGUA
SHOWER

"Lota Agua" is a low-cost folding shower for use in areas where there is no running water available. The user simply stands on a container filled with water. The water pressure required is created by the user's body weight – a valve on the shower head controls the water flow. The shower is very easy to set up. The water used for showering can then be directly recycled for watering gardens and plants.

"A concept that focuses on the problem of keeping clean in disaster-struck regions. Low-tech – very easy to set up on location. A good solution for transport and the easy creation of water pressure using the user's body weight have us convinced!"

"I am absolutely honored to receive this award as a Canadian designer and hope for more countries around the world to create award-winning designs recognized by iF."

iF INDUSTRIE FORUM DESIGN E.V.

LIST OF MEMBERS AND ARTICLE OF ASSOCIATION

BOARD

Chairman

Ernst Raue
Deutsche Messe AG
Messegelände
30521 Hannover
Germany
www.messe.de

Christoph Böninger
Böcklinstraße 54
80638 München
Germany
www.auerberg.eu

Prof. Fritz Frenkler
f/p design GmbH
Mauerkicherstraße 4
81679 München
Germany
www.fp-design-gmbh.com

Boris Kochan
Kochan & Partner GmbH
Werbung, Design und Kommunikation
Hirschgartenallee 25
80639 München
Germany
www.kochan.de

Susanne Schmidhuber
Schmidhuber + Partner GbR
Architektur Innenarchitektur
Nederlinger Straße 21
80638 München
Germany
www.schmidhuber.de

Managing Director

Ralph Wiegmann
iF Industrie Forum Design e.V.
Bahnhofstrasse 8
30159 Hannover
Germany
www.ifdesign.de

Honorary Members

Knut Bliesener
Im Haspelfelde 31
30173 Hannover
Germany

Sepp D. Heckmann
Schopenhauerstr. 12
30625 Hannover
Germany

Prof. Herbert H.Lindinger
Brahmsstraße 3
30177 Hannover
Germany
www.lindingerdesign.de

Herbert H. Schultes
Rodelbahnstraße 1
82256 Fürstenfeldbruck
Germany

Corporate members

AL-KO Geräte GmbH
Ichenhauser Straße 14
89359 Kötz
Germany
www.al-ko.de

Artemide GmbH
Hans-Böckler-Straße 2
58730 Fröndenberg
Germany
www.artemide.de

B / S / H Bosch und Siemens Hausgeräte GmbH
Carl-Wery-Straße 34
81793 München
Germany
www.bsh-group.de

BASF SE
Designfabrik E-KTE/IF - F206, Zi. 563
E-KTE/IF - F 206, Zi. 563
67056 Ludwigshafen
Germany
www.basf.com

BEGA Gantenbrink-Leuchten
GmbH + Co
Hennenbusch 1
58708 Menden
Germany
www.bega.com

Blomus GmbH
Zur Hubertushalle 4
59846 Sundern
Germany
www.blomus.com

BMW AG
Knorrstraße 147
80788 München
Germany
www.bmwgroup.com

BREE Collection GmbH & Co. KG
Gerberstraße 3
30916 Isernhagen
Germany
www.bree.de

Carpet Concept
Objekt-Teppichboden GmbH
Bunzlauer Straße 7
33719 Bielefeld
Germany
www.carpet-concept.de

CLAGE GmbH
Pirolweg 1-5
21337 Lüneburg
Germany
www.clage.de

Crown Gabelstapler GmbH & Co. KG
Moosacher Straße 52
80809 München
Germany
www.crown.com/de

Daimler AG
Mercedes-Benz Design
Werk 059, HPC X800
71059 Sindelfingen
Germany
www.daimler.com

design report - Blue C. Verlag
Ein Unternehmen der Konradin Medien GmbH
Ernst-Mey-Straße 8
70771 Leinfelden-Echterdingen
Germany
www.design-report.de

designaffairs GmbH
Rosenheimer Straße 145 b / EG
81671 München
Germany
www.designaffairs.com

Deutsche Messe AG
Messegelände
30521 Hannover
Germany
www.messe.de

Die Neue Sammlung, Pinakothek der Moderne
Türkenstraße 15
80333 München
Germany
www.die-neue-sammlung.de

EnBW AG
Messen und Events
Obere Stegwiesen 29
88400 Biberach
Germany
www.enbw.com

EnBW AG
Marketing und Koordination
Durlacher Allee 93
76131 Karlsruhe
Germany
www.enbw.com

Expotechnik Heinz Soschinski GmbH
Aarstraße 176
65232 Taunusstein
Germany
www.expotechnik.com

F. W. OVENTROP GmbH & Co. KG
Paul-Oventrop-Straße 1
59939 Olsberg
Germany
www.oventrop.de

FESTO AG & Co. KG
Leiter Design Coordination
Ruiter Straße 82
73734 Esslingen
Germany
www.festo.com

GHM Gesellschaft für Handwerksmessen mbH
Willy-Brandt-Allee 1
81829 München
Germany
www.ghm.de

Gira Giersiepen GmbH & Co. KG
Dahlienstraße
42477 Radevormwald
Germany
www.gira.de

Grohe AG
Design Studio
Feldmühleplatz 15
40545 Düsseldorf
Germany
www.grohe.de

häfelinger + wagner design gmbh
Türkenstraße 55-57
80799 München
Germany
www.hwdesign.de

Hansgrohe AG
Auestraße 5-9
77761 Schiltach
Germany
www.hansgrohe.com

HBK Braunschweig
Johannes-Selenka-Platz 1
38118 Braunschweig
Germany
www.hbk-bs.de

HEWI Heinrich Wilke GmbH
Postfach 1260
34442 Bad Arolsen
Germany
www.hewi.de

Hiller Objektmöbel GmbH & Co. KG
Kippenheimer Straße 6
77971 Kippenheim
Germany
www.hiller-moebel.de

IBM Deutschland GmbH
IBM-Allee 1
71139 Ehningen
Germany
www.ibm.com/de

interstil - Diedrichsen GmbH
Liebigstraße 1-3
33803 Steinhagen
Germany
www.interstil.de

Interstuhl Büromöbel GmbH & Co. KG
Brühlstraße 21
72469 Meßstetten-Tieringen
Germany
www.interstuhl.de

Isaria Corporate Design AG
Gewerbepark Aich 7-9
85667 Oberpframmern
Germany
www.isaria.com

JAB Teppiche Heinz Anstoetz KG
Dammheider Straße 67
32052 Herford
Germany
www.jab.de

Kermi GmbH
Pankofen-Bahnhof 1
94447 Plattling
Germany
www.kermi.de

Kochan & Partner GmbH
Hirschgartenallee 25
80639 München
Germany
www.kochan.de

Köttermann GmbH & Co. KG
Industriestraße 2-10
31311 Uetze
Germany
www.kottermann.com

KWC AG
Hauptstraße 57
5726 Unterkulm
Switzerland
www.kwc.ch

LOEWE OPTA GmbH
Leiter Design/Designmanagement
Industriestraße 11
96317 Kronach
Germany
www.loewe.de

MeisterSinger GmbH & Co. KG
Hafenweg 46
48155 Münster
Germany
www.meistersinger.de

Merten GmbH
Fritz-Kotz-Straße 8
51674 Wiehl
Germany
www.merten.de

MüllerKälber GmbH
Daimlerstraße 2
71546 Aspach
Germany
www.muellerkaelber.de

Niedersächsisches Ministerium für Wirtschaft,
Technologie u. Verkehr
Friedrichswall 1
30159 Hannover
Germany
www.mw.niedersachsen.de

Nils Holger Moormann GmbH
An der Festhalle 2
83229 Aschau
Germany
www.moormann.de

OCTANORM-Vertriebs-GmbH
für Bauelemente
Raiffeisenstraße 39
70794 Filderstadt
Germany
www.octanorm.de

PH MEYER Wirtschaftsberatung GmbH & Co. KG
Mittelstraße 50
33602 Bielefeld
Germany
www.phmeyer.de

Philips GmbH
UB Consumer Lifestyle
Lübeckertordamm 5
20099 Hamburg
Germany
www.philips.de

PLANMECA OY
Asentajankatu 6
00880 Helsinki
Finland
www.planmeca.com

Poggenpohl Möbelwerke GmbH
Poggenpohl Straße 1
32051 Herford
Germany
www.poggenpohl.de

REALTIME TECHNOLOGY (RTT)
Rosenheimer Straße 145
81671 München
Germany
www.rtt.ag

RITTO GmbH
Rodenbacher Straße 15
35708 Haiger
Germany
www.ritto.de

Schendel & Pawlaczyk Messebau GmbH
Im Derdel 3
48167 Münster
Germany
www.schendel-pawlaczyk.de

SCHULTE Duschkabinenbau KG
Zum Dümpel 1
59846 Sundern
Germany
www.schulte.de

Sedus Stoll AG
Brückenstraße 15
79761 Waldshut-Tiengen
Germany
www.sedus.de

Seibel Designpartner GmbH
Industriestraße 5
40822 Mettmann
Germany
www.seibel-designpartner.de

Sennheiser electronic GmbH & Co. KG
Am Labor 1
30900 Wedemark
Germany
www.sennheiser.com

Siemens AG
Abt. CC MC1
Wittelsbacherplatz 2
80333 München
Germany
www.siemens.com

SKYLOTEC GmbH
Im Bruch 11-15
56567 Neuwied
Germany
www.skylotec.de

SMA Solar Technology AG
Sonnenallee 1
34266 Niestetal
Germany
www.sma.de

Steelcase Werndl AG
Georg-Aicher-Straße 7
83026 Rosenheim
Germany
www.steelcase-werndl.de

Storck Bicycle GmbH
Rudolfstraße 1
65510 Idstein
Germany
www.storck-bicycle.de

TRILUX GmbH + Co KG
Heidestraße 4
59759 Arnsberg
Germany
www.trilux.de

Vauth-Sagel
Systemtechnik GmbH & Co. KG
Neue Straße 27
33034 Brakel-Erkeln
Germany
www.vauth-sagel.de

Viessmann Werke GmbH Co. KG
Viessmannstraße 1
35108 Allendorf/Eder
Germany
www.viessmann.de

Volkswagen AG
Volkswagen Design
Brieffach 1701, Berliner Ring 2
38436 Wolfsburg
Germany
www.volkswagen.de

wiege Entwicklungs GmbH
Hauptstraße 81
31848 Bad Münder
Germany
www.wiege.com

Wilkhahn
Wilkening + Hahne GmbH + Co. KG.
Fritz Hahne Straße 8
31848 Bad Münder
Germany
www.wilkhahn.de

WINI Büromöbel GmbH & Co. KG
Auhagenstraße 79
31863 Coppenbrügge
Germany
www.wini.de

wodtke GmbH
Rittweg 55-57
72070 Tübingen-Hirschau
Germany
www.wodtke.com

Yokogawa Electric Corporation
2-9-32 naka-cho
180-8750 Mushashino-shi, Tokyo
Japan
www.yokogawa.com

Zumtobel Lighting GmbH
Schweizer Straße 30
6850 Dornbirn
Austria
www.zumtobel.de

DESIGN STUDIOS OR DESIGNER

3-point concepts GmbH
Reichenberger Straße 124
10999 Berlin
Germany
www.3pc.de

aka buna design consult
Nonnbergstiege 1
5020 Salzburg
Austria
www.akabuna.at

ArGe Design braucht Täter
c/o id 3 Michael Gehlen
Ägidiusstraße 1
50937 Köln
Germany
www.martin-neuhaus.com

artcollin
Goethestraße 28
80336 München
Germany
www.artcollin.de

ArteFakt Pohl Fiegl
Liebigstraße 50 - 52
64293 Darmstadt
Germany
www.artefakt.de

at-design
Flugplatzstraße 111
90768 Fürth
Germany
www.atdesign.de

Avidus Consulting
Wielandstraße 1
65187 Wiesbaden
Germany
www.avidus-consulting.com

bayern design forum e.V.
Luitpoldstraße 3
90402 Nürnberg
Germany
www.bayern-design.de

B:SiGN Werbeagentur GmbH
Ellernstraße 36
30175 Hannover
Germany
www.bsign.de

Braake Design
Turnierstraße 3
70599 Stuttgart
Germany
www.braake.com

brodbeck design
Schillerstraße 40 c
80336 München
Germany
www.brodbeckdesign.de

D´ART Visuelle Kommunikation GmbH
Adlerstraße 41
70199 Stuttgart
Germany
www.dartwork.de

design studio hartmut s. engel
Monreposstraße 7
71634 Ludwigsburg
Germany

Design Tech
Zeppelinstraße 53
72119 Ammerbuch
Germany
www.designtechschmid.de

Deutscher Designer Club e.V.
Große Fischestraße 7
60311 Frankfurt am Main
Germany
www.ddc.de

DRWA Das Rudel Werbeagentur
Erbprinzenstraße 11
79098 Freiburg
Germany
www.drwa.de

f/p design GmbH
Mauerkicherstraße 4
81679 München
Germany
www.f-p-design.com

FRACKENPOHL POULHEIM GMBH
Luxemburger Str. 72
50674 Köln
Germany
www.frackenpohl-poulheim.de

Fuenfwerken Design AG
Taunusstraße 52
65183 Wiesbaden
Germany
www.fuenfwerken.com

GDC-Design
Ludwigstraße 2
90763 Fürth
Germany
www.gdc-design.de

GUTE GESELLSCHAFT FÜR STRATEGIE, DESIGN UND KOMMUNIKATION MBH
Grafenberger Allee 126
40237 Düsseldorf
Germany
www.gutegesellschaft.com

H H Schultes Design Studio
Rodelbahnstraße 1
82256 Fürstenfeldbruck
Germany

hymer idc GmbH + Co. KG
Innovations- u. Design Center
Blücherstraße 32
75177 Pforzheim
Germany
www.hymer-idc.de

i / i / d Institut für Integriertes Design
an der HfK Bremen
Am Speicher XI, Abtlg.7, Boden 3
28217 Bremen
Germany
www.iidbremen.de

identis GmbH
Bötzinger Straße 36
79111 Freiburg
Germany
www.identis.de

INOID DesignGroup
Reutlingerstraße 114
70597 Stuttgart
Germany
www.inoid.de

Interbrand
Kirchenweg 5
8008 Zürich
Switzerland
www.interbrand.com

MEDIADESIGN HOCHSCHULE
für Design und Informatik
Claudius-Keller-Straße 7
81669 München
Germany
www.mediadesign.de

molldesign
Turmgasse 7
73525 Schwäbisch Gmünd
Germany
www.molldesign.de

mormedi
Plaza Republica Argentina 3
28002 Madrid
Spain
www.momedi.com

Nova Design Co., Ltd.
Tower C, 8F, No. 96, Sec. 1, Xintai 5th Rd.
221 Taipei Country
Taiwan
www.e-novadesign.com

OCO-Design O. K. Nüsse
An der Kleimannbrücke 79
48157 Münster
Germany
www.oco-design.de

Olaf Hoffmann Industrial Design
Metzstraße 14 b
81667 München
Germany
www.olaf-hoffmann-design.de

Oldenbourg Industrieverlag GmbH
Rosenheimer Str. 145
81671 München
Germany
www.gwf-wasser-abwasser.de

Philips International BV Philips Design
Building HWD, Emmasingel 24
5600 Eindhoven
The Netherlands
www.philips.nl

Pilotfish GmbH
Schleissheimer Straße 6
80333 München
Germany
www.pilotfish.eu

Polvan Design Ltd.
Cemil Topuzlu cad. 79/2 Caddebostan
34170 Istanbul
Turkey
www.polvandesign.com

PROMOTIONAL iDEAS Werbeagentur GmbH
Hessenring 76
61348 Bad Homburg
Germany
www.promotionalideas.de

rahe+rahe design
Konsul-Smidt-Straße 8c
28217 Bremen
Germany
www.rahedesign.de

Rokitta Produkt & Markenästhetik
Kölner Straße 38 a
45481 Mühlheim an der Ruhr
Germany
www.rokitta.de

S + F Architektur + Design
J. Schwedux + R. Filipschack
Straße der Nationen 5, Expo-Park Ost
30539 Hannover
Germany
www.schwedux-filipschack.de

SCHOLZ & VOLKMER GmbH
Schwalbacher Straße 72
65183 Wiesbaden
Germany
www.s-v.de

Spirit Design Consulting & Services GmbH
Silbergasse 8
1190 Wien
Austria
www.spiritdesign.com

Strategy&Marketing Institute GmbH
Lange-Hop-Straße 19
30559 Hannover
Germany

Taipei Base Design Center
1F.,No.4,Ln.176,Sec.1,Da'an Rd.,Da'an Dist.
106 Taipei City
Taiwan
www.asia-bdc.com

TRICON Design AG
Bahnhofstraße 26
72138 Kirchentellinsfurt
Germany
www.tricon-design.de

Ueberholz GmbH
Warndtstraße 7
42285 Wuppertal
Germany
www.ueberholz.de

UNIPLAN GmbH & Co. KG
Schanzenstraße 39 a/b
51063 Köln
Germany
www.uniplan.com

VDID/DDV
Markgrafenstraße 15
10969 Berlin
Germany
www.vdid.de

Weinberg & Ruf
Produktgestaltung
Martinsstraße 5
70794 Filderstadt
Germany
www.weinberg-ruf.de

INDIVIDUAL MEMBERS

Thomas Bade
Im Knick 9
31655 Stadthagen
Germany

Olaf Barski
Hermannstraße 15
60318 Frankfurt
Germany
www.barskidesign.com

Thomas Biswanger
Probierlweg 47
85049 Ingolstadt
Germany
www.thomasbiswanger.de

Christoph Böninger
Auerberg 1
83730 Fischbachau
Germany
www.auerberg.eu

Gerd Bulthaup
Chamissostraße 1
81925 München
Germany

Nenad Dordevic
Türkenstraße 103
80799 München
Germany
www.nexus-group.de

Prof. Dr. Gerdum Enders
Terrasse 19
34117 Kassel
Germany
www.globalmind.net

Simon Philipp Engels
Bergstraße 2
59846 Sundern
Germany

Christoph Eschke
Admiralitätstraße 10
20459 Hamburg
Germany
www.ebnerstolz.de

Andreas Gantenhammer
Meerbuscher Straße 64-78
40670 Meerbusch
Germany
www.gantenhammer.de

Michael Grüter
Hainholzstraße 17
31558 Hagenburg
Germany

Peter Hartmann
Fasaneriestraße 10
80636 München
Germany
www.h-bc.de

Josef Hasberg
Lokenbach 8 - 10
51491 Overath
Germany

Bibs Hosak-Robb
Mendelssohnstraße 31
81245 München
Germany
www.bibs-design.de

K. Michael Kühne
Kampstraße 96
22113 Oststeinbeck
Germany
www.creacon.de

Sebastian Le Peetz
Osterstraße 43 a
30159 Hannover
Germany

Hildegund Lichtwark
Rudolf-Diesel-Straße 4
41516 Grevenbroich
Germany

Johannes Loer
Luisenstraße 24
59379 Selm-Bork
Germany
www.metamorphdesignwerk.de

Kai Miethig
P.O. Box 11224, Wind Tower, 6th Floor
Diplomatic Area, Manama
Bahrain
www.bg-ca.com

Seyedeh Fatemeh Neseriniaki
Di San Bernardo 8/5
16123 Genova
Italy

Hamidreza Nikkhou
#2.No79. Shahrak-Baharan kashani St.
1487833-455 Teheran
Iran

Dr. Günther Portenlänger
Wolfrathshauser Straße 77
82049 Pullach
Germany

Seyed Mansour Pour Mohseni Shakib
No.281- Yasseman Alley-Manzariyeh-Namjo St.
41936 Rasht
Iran

Seyed Mahdi Pour Mohseni Shakib
No.281- Yasseman Alley-Manzariyeh-Namjo St.
41936 Rasht
Iran

Christoph Rohrer
Roecklplatz 3
80469 München
Germany
www.kms-team.de

Karina Rudolph
Altes Dorf 6a
33699 Bielefeld
Germany
www.schoenformerei.de

HD Schellnack
Mintropstrasse 61
45329 Essen
Germany
www.nodesign.com

Eberhard Schlegel
Am Kapellenweg 4
88525 Dürmentingen
Germany
www.schlegel.biz

Prof. Gunnar Spellmeyer
c/o Fachhochschule Hannover Expo Plaza 2
30539 Hannover
Germany
www.mixxd.com

Andreas Thierry
Arthur-Kutscher-Platz 1 / VII
80802 München
Germany

Peter Thonet
Michael-Thonet-Str. 1
35066 Frankenberg
Germany
www.thonet.de

Prof. Martin Topel
Fuhlrottstraße 10, Gebäude I, Ebene 16, Raum 76
42119 Wuppertal
Germany
www.visionlabs.org

Roland Wagner
Tullstraße 19
77933 Lahr
Germany
www.wagner-system.de

Gabriel Weber
Orffstraße 35
80637 München
Germany
www.rausweber.de

ARTICLES OF ASSOCIATION OF "iF – INDUSTRIE FORUM DESIGN E. V."

ARTICLE 1. NAME AND REGISTERED OFFICE OF THE ASSOCIATION

1. *The name of the Association is "iF – Industrie Forum Design e. V.".*
2. *iF – Industrie Forum Design is listed in the Register of Associations (Vereinsregister).*
3. *The Association's registered office is in Hannover, Germany.*

ARTICLE 2. OBJECT OF THE ASSOCIATION

The object of the Association is to support and gain acceptance for design as a link in the value chain and as a cultural component of society. The Association considers the targeted design of products, public and private living spaces, and user-friendly software applications as constituting its own civic and cultural mission. The Association aims to:

1. *Recognize design achievements that help companies achieve their business goals and cement their economic success*
2. *Organize competitions, exhibitions, conferences, lectures, and other events*
3. *Issue publications to serve as the basis for discussions*
4. *Bolster the public's awareness of design*
5. *Provide a forum for objective dialogue on design-related issues.*

The Association's activities give it a regional, national, and international presence.

ARTICLE 3. FINANCIAL YEAR

The financial year corresponds to the calendar year.

ARTICLE 4. MEMBERSHIP

1. *Any natural person or legal entity may become a member of the Association. Executive Management shall examine and decide on membership applications, which are to be submitted in writing. In cases where Executive Management is unsure, the Board of Management shall decide on whether to grant membership by means of a simple majority vote. The Association is not required to justify a negative decision on membership.*
2. *In addition to members, the Association also consists of honorary members. Honorary members are individuals who have been particularly meritorious in their support of "iF – Industrie Forum Design". They may be elected to honorary membership by way of a unanimous resolution passed at the General Meeting. They are not entitled to vote at the General Meeting.*
3. *Membership shall end*
   a) Through resignation, which must be declared to the Association in writing. This declaration requires a minimum of three months' advance notice before the end of the financial year, and takes effect at the end of the financial year.
   b) By way of expulsion upon a unanimous resolution by the Board of Management.
   c) At the time of a member's death.

ARTICLE 5. OFFICIAL BODIES

The Association's official bodies are:

1. *The General Meeting*
2. *The Board of Management*
3. *Executive Management.*

ARTICLE 6. GENERAL MEETING

1. *The General Meeting is responsible for:*
   a) Receiving the reports of the Board of Management and the independent auditors
   b) Electing the Board of Management
   c) Approving the actions of the Board of Management and the Executive Management
   d) Changing the Articles of Association
   e) Dissolving the Association
   f) Using the Association's assets in the event that it should be dissolved
   g) Using the annual profits or handling of any loss
   h) Setting the annual membership fee
   i) Electing honorary members
2. *A regular General Meeting will be held annually. All members of the Association shall be invited to attend the General Meeting. The agenda of the regular General Meeting will, in particular, include reports on the preceding financial year, on the Association's financial statements, and on the results of the auditing of accounts.*
3. *General Meeting shall be called in writing by the Chairperson of the Board of Management, with a minimum of one week's advance notice and with such notice containing a meeting agenda. The agenda of the General Meeting is set by the Chairperson of the Board of Management or by another member of the Board of Management appointed by the Chairperson. Topics requested by individual members may be added to the agenda if the General Meeting gives its unanimous consent.*
4. *The Chairperson of the Board of Management may call an extraordinary General Meeting if an important reason exists for doing so. A meeting of this kind must be called if it has been requested in writing by at least one third of all members. Two day's advance notice is required. In urgent cases, invitations may be transmitted by telephone or fax.*
5. *General Meetings shall be presided over by the Chairperson of the Board of Management or, in the event that he or she should be prevented from doing so, by another member of the Board of Management appointed by the Chairperson.*
6. *A General Meeting that has been properly called has a quorum if, in addition to the individual presiding over the Meeting, at least three other members in possession of voting rights are present.*

7. *Each member has one vote at the General Meeting. Unless otherwise stipulated in these Articles, resolutions of the General Meeting shall be passed by way of a simple majority of those members present and entitled to vote. Passing a resolution on the appropriation of profit carried forward or the treatment of a loss requires a majority of 80 percent among the voting members present. Members entitled to vote may appoint a proxy to attend the General Meeting in their place, including a proxy voting right. In the of a tie, the deciding vote shall be cast by the individual presiding over the Meeting.*
8. *Urgent resolutions may be passed by round-robin procedure.*
9. *Resolutions on amendments to the Articles of Association shall require a three-quarters majority vote by the attending members who are eligible to vote.*
10. *Approved resolutions must be recorded in writing and signed by both the individual presiding over the Meeting and the keeper of the minutes. All members shall receive a copy of the minutes of the General Meeting.*

ARTICLE 7. BOARD OF MANAGEMENT

1. *The Association's Board of Management consists of the Chairperson and a maximum of six other members. The members of the Board of Management will be elected by the members of the Association for a term of four years. They shall remain in office until new elections take place. Elections may be revoked at General Meetings for good cause.*
2. *The Association will be represented in and out of court by the Chairperson of the Board of Management, who shall be permitted to delegate a power of attorney to third persons in individual cases.*
3. *The Board of Management shall support Executive Management in the running of the Association and shall decide on all matters concerning the Association, providing that the right to such decision-making is not reserved by the General Meeting. The Board of Management may elect other individuals required to assume administrative tasks and will decide on all of the Association's income and expenditures.*
4. *At the General Meeting, the members will decide on special activities and related expenditures as presented by the Board of Management.*
5. *The General Meeting may provide the Board of Management with rules of procedure governing the Association's operation.*

ARTICLE 8. EXECUTIVE MANAGEMENT

The Board of Management shall appoint the executive management, which will be charged with executing the ongoing tasks of the Association in accordance with the guidelines issued by the Board of Management. Executive Management is authorized to make decisions on membership applications for new members. Executive Management shall inform the Board of Management about all significant matters affecting the Association. Executive Management shall follow all instructions issued by the Board of Management and shall consult the Board of Management in advance regarding all important matters.

ARTICLE 9. MEMBERSHIP FEES

The annual membership fee will be set at the General Meeting and is currently as follows:

– Companies:	*600 Euro minimum*
– Designers and design studios:	*300 Euro minimum*
– Individuals:	*150 Euro minimum.*

ARTICLE 10. AUDITING

Accounting control will be assigned to Deutsche Messe AG's auditing department or to an independent auditor. The results of auditing will be reported at the regular General Meeting.

ARTICLE 11. DISSOLUTION OF THE ASSOCIATION

1. *A resolution to dissolve the Association shall require a three-fourths majority vote by the members who are entitled to vote, provided that at least two thirds of the members entitled to vote are present at the General Meeting.*
2. *Upon dissolution or annulment of the Association or in the event that the current object of the Association should cease to be applicable, any assets exceeding the capital shares deposited by the members and the fair market value of non-cash capital contributions made by members shall go to a public corporation that will use the assets for cultural or community welfare purposes. Resolutions on the future application of such assets may only be acted upon subsequent to consent thereto by the tax authorities.*
3. *The General Meeting which resolves to dissolve the Association shall pass a resolution on the application of the Association's assets by way of a three-fourths majority vote.*

ARTICLE 12. PROFITS, EXPENDITURES, ACCOUNTING

1. *Any profits produced by the Association shall be used exclusively for the tax-privileged purposes of these Articles of Association. Members shall not receive shares of profit, nor shall they, in their capacity as members, receive any other forms of bestowal from Association funds.*
2. *No individual may be favored by receiving disproportionately high remuneration for administrative tasks not in keeping with the object of the Association.*
3. *Proof that the Association's funds are being used properly shall be guaranteed by way of adequate and orderly accounting procedures.*

Hannover, June 5, 2008

Ernst Raue
Chairman

iF PARTNER AND SPONSORS 2012

iF official partner

AGD — Allianz deutscher Designer
Germany

AXIS — AXIS – the leading design magazine
Japan

BCD Barcelona Centre de Disseny — Barcelona Design Centre
Spain

BNO — Beroepsorganisatie Nederlandse
Ontwerpers, Netherlands

BDIA Bund Deutscher Innen Architekten — Bund Deutscher Innen Architekten
Germany

B V D W Bundesverband Digitale Wirtschaft — Bundesverband Digitale Wirtschaft
Germany

CIDA Taiwan — China Industrial Designers Association
Taiwan

designaffairs — designaffairs GmbH
Germany

designaustria — Design Austria e. V.
Austria

DDC — Deutscher Designer Club e. V.
Germany

Fraunhofer — Fraunhofer Gesellschaft
Germany

HAUTE INNOVATION MATERIAL AND TECHNOLOGY — Haute Innovation
Material and Technology
Germany

IIDA Israel Industrial Designers Association — Israel Industrial Designers Association
Israel

IdcN International Design Center NAGOYA — International Design Center
NAGOYA Inc.
Japan

material sense — Material Sense
Institute for Materials Innovation
Netherlands

HUNGARIAN DESIGN COUNCIL mft — MFT – Hungarian Design Council
Hungary

राष्ट्रीय डिज़ाइन संस्थान NATIONAL INSTITUTE OF DESIGN — National Institute of Design India
India

nordmedia — Die Mediengesellschaft
Niedersachsen/Bremen mbH
Germany

PAGE — das führende design- und publishing
magazin, Germany

www.packagingconnections.com — Packaging Connections
Providing Online Packaging Services and
Packaging Consultancy Worldwide
India

raumPROBE
Germany

sda – swiss design association
Switzerland

Swiss Graphic Designers
Switzerland

Association of Applied Graphic Designers
Poland

Taipei Design Center Düsseldorf
Germany

Taiwan Design Center
Taiwan

VDID / DDV
Germany

Wochenmagazin für Marketing, Werbung, Medien und E-Business, Germany

zwomp.de – Das Online-Magazin für Industriedesign und Innovation, Germany

Supporter iF DESIGN TALENTS

LG Electronics Ltd.
Korea

GOLD Sponsors iF concept design award

CeBIT
Germany

Hansgrohe AG
Germany

iF International Forum Design GmbH
Germany

LG Hausys Ltd.
Korea

Samsung Electronics
Korea

Volkswagen AG
Germany

BRONZE Sponsor iF concept design award

Woongjin Coway
Korea

Partner iF concept design award

designxport GmbH
Germany

Media partner iF concept design award

Yanko Design
Canada

INDEX PARTICIPANTS

iF RANKING

The iF ranking university is only based on the winning entries of the student-focused "iF concept design award". When a young designer wins an "iF concept design award", his or her university receives 20 points. If a university has more than one successful participant, the points add up, and the ranking position of the university becomes higher. The results of the last three years iF concept design awards compose all current ranking positions. Each new ranking is published after the yearly awards ceremony.

Universities from around the world can compare themselves with other institutions of higher learning in this iF ranking university. It is also available online: www.ifdesign.de/talents_universities_e.

The position is displayed in the first column; the points achieved by students of the universities are visible in the second column. Universities who have achieved the same amount of points share the same position. For example three Universities have 120 points and thus are all position 4 in the ranking. Still the numbers of positions add up, which is why the next position after 4 is 6.

iF ranking university

Position	Points	University	Country
1	360	National Taiwan University of Science and Technology	Taiwan
2	220	Samsung Art & Design Institute	South Korea
2	220	Zhejiang University	China
4	160	Fachhochschule Hannover	Germany
4	160	Umeå Universitet	Sweden
6	140	Shih Chien University, Taipei	Taiwan
7	120	Jiangnan University / School of Design, Wuxi	China
7	120	National Cheng Kung University	Taiwan
7	120	National Taiwan University	Taiwan
7	120	Sangmyung University	South Korea
11	100	Dalian Nationalities University	China
11	100	Hongik University	South Korea
11	100	Konkuk University	South Korea
11	100	Korea Advanced Institute of Science and Technology - KAIST	South Korea
11	100	National Taipei University of Technology	Taiwan
16	80	Chang Gung University	Taiwan
16	80	Hochschule für Gestaltung Schwäbisch Gmünd	Germany
16	80	Keimyung University	South Korea
16	80	Royal College of Art, London	United Kingdom
16	80	Seoul National University of Science and Technology	South Korea
16	80	Shu-Te University	Taiwan
22	60	Burg Giebichenstein Hochschule für Kunst und Design, Halle	Germany
22	60	Daegu University	South Korea
22	60	Fachhochschule München	Germany
22	60	Hochschule Coburg	Germany
22	60	Hochschule für Gestaltung Offenbach am Main	Germany
22	60	Huazhong University of Science & Technology	China
22	60	Kookmin University	South Korea
22	60	Ming-Chi University of Technology	Taiwan
22	60	National Taiwan Normal University (NTNU)	Taiwan
22	60	Tatung University, Taipei	Taiwan
32	40	Art Center College of Design	Taiwan
32	40	Bergische Universität Wuppertal	Germany
32	40	Cheongju University	South Korea
32	40	China Academy of Art	China
32	40	Chonbuk National University, Jeonju	South Korea
32	40	Dankook University	South Korea

Position	Points	University	Country
32	40	Fachhochschule Aachen	Germany
32	40	Fachhochschule Dortmund	Germany
32	40	Hanyang University	South Korea
32	40	Hochschule für Gestaltung u. Kunst Basel FHNW	Switzerland
32	40	Hochschule für Technik und Wirtschaft Dresden	Germany
32	40	Korea University	South Korea
32	40	Kunsthochschule Berlin-Weißensee	Germany
32	40	Lund University	Sweden
32	40	Muthesius Kunsthochschule, Kiel	Germany
32	40	Nanchang University	China
32	40	National Yunlin University of Science and Technology	Taiwan
32	40	Pontifícia Universidade Católica do Paraná	Brazil
32	40	Rhode Island School of Design	United States of America
32	40	Rochester Institute of Technology	United States of America
32	40	Seoul National University	South Korea
32	40	Seoul Women's University	South Korea
32	40	Sookmyung Women's University	South Korea
32	40	Tainan University of Technology	Taiwan
32	40	Tianjin Academy of Fine Arts	China
32	40	Tsinghua University	China
32	40	Universität Duisburg-Essen	Germany
32	40	Xi'an University of Science & Technology	China
60	20	Academy of Art University	United States of America
60	20	Art Institute of Chicago	United States of America
60	20	Asia University	Taiwan
60	20	Azad University of Tehran	Islamic Republic of Iran
60	20	Beijing Institute of Fashion Technology	China
60	20	Beijing Institute of Technology	China
60	20	Beijing Technology and Business University	China
60	20	Bucheon University	South Korea
60	20	Catholic University of Daegu	South Korea
60	20	Centro Universitário Belas Artes São Paulo	Brazil
60	20	Changwon National University	South Korea
60	20	Concordia University	Canada
60	20	Dalian Polytechnic University	China
60	20	DA-YEH university	Taiwan
60	20	Designskole Kolding	Denmark

iF ranking university

Position	Points	University	Country
60	20	Domus Academy	Italy
60	20	Dong-A University	South Korea
60	20	East China Normal University	China
60	20	Emily Carr University	Canada
60	20	Fachhochschule Trier	Germany
60	20	Fachhochschule Würzburg-Schweinfurt	Germany
60	20	FHNW Aarau	Switzerland
60	20	Folkwang Hochschule	Germany
60	20	Freie Universität Bozen	Italy
60	20	Georg-Simon-Ohm Hochschule Nürnberg	Germany
60	20	Guangdong University of Technology	China
60	20	Handong University	South Korea
60	20	Hannam University	South Korea
60	20	Hochschule für Bildende Künste Braunschweig	Germany
60	20	Hochschule für bildende Künste Hamburg	Germany
60	20	Hochschule Magdeburg-Stendal (FH)	Germany
60	20	Hochschule Niederrhein, Krefeld + Mönchengladbach	Germany
60	20	Hogeschool Utrecht	Netherlands
60	20	Hohai University	China
60	20	Hoseo Unviersity	South Korea
60	20	Hubei University of Technology	China
60	20	Hunan University	China
60	20	Hyupsung University	South Korea
60	20	Inha University	South Korea
60	20	Inje University	South Korea
60	20	Institute of Design (IN.D.) Düsseldorf	Germany
60	20	Institut Teknologi Sepuluh Nopember	Indonesia
60	20	Jeju National University	South Korea
60	20	Jiangxi Universit of Finance and Economics	China
60	20	Kongju National University	South Korea
60	20	Korea Polytechnic University	South Korea
60	20	Kunsthochschule für Medien Köln	Germany
60	20	Kyoto Institute of Technology	Japan
60	20	Kyungwon University	South Korea
60	20	Ling-Tung University	Taiwan
60	20	Luxun Academy of Fine Arts	China
60	20	Miami Ad School Europe	United States of America

Position	Points	University	Country
60	20	Ming Chuan University	Taiwan
60	20	Nan Hua University	Taiwan
60	20	Nanjing Forestry University	China
60	20	National Chengchi University	Taiwan
60	20	National Chiao Tung University	Taiwan
60	20	National Kaohsiung Normal University	Taiwan
60	20	National United University	Taiwan
60	20	North Carolina (NC) State University	United States of America
60	20	Scuola Politecnica di Design	Italy
60	20	Shanghai Institute of Visual Art	China
60	20	Shenkar College of Engineering and Design	Israel
60	20	Shenyang Jianzhu University	China
60	20	South China Agricultural University	China
60	20	South China University of Technology	China
60	20	Southern Taiwan University of Technology	Taiwan
60	20	Strate Collège, Paris	France
60	20	Tianjin University of Science and Technology	China
60	20	Universidade do Estado de Minas Gerais	Brazil
60	20	Universidade Federal Campina Grande	Brazil
60	20	Universidade Federal do Rio Grande do Sul-UFRGS	Brazil
60	20	University of Ljubljana - Academy of Fine Arts and Design	Slovenia
60	20	University Of Ulsan	South Korea
60	20	Vilnius Academy of Arts	Lithuania
60	20	Yonsei University	South Korea
60	20	Zhejiang Gongshang University	China

INDEX UNIVERSITIES

Index of universities

iFDESIGN.DE

SERVING PROFESSIONALS.
SUPPORTING STUDENTS.

iF DESIGN AWARDS.
GO FOR THE BEST.

iFDESIGN.DE

BORN IN GERMANY
HOME WORLDWIDE.